Archibald Geikie

Geological Sketches at Home and Abroad

Archibald Geikie

Geological Sketches at Home and Abroad

ISBN/EAN: 9783744678865

Printed in Europe, USA, Canada, Australia, Japan

Cover: Foto ©ninafisch / pixelio.de

More available books at **www.hansebooks.com**

GEOLOGICAL SKETCHES

AT HOME AND ABROAD

BY

ARCHIBALD GEIKIE, LL.D., F.R.S.,
DIRECTOR GENERAL OF THE GEOLOGICAL SURVEYS OF THE UNITED KINGDOM

WITH ILLUSTRATIONS

New York
MACMILLAN AND CO.
1892

PREFACE.

THE Essays here collected and revised have appeared at intervals in various journals, and I have to express my thanks to the Councils of the Royal Society of Edinburgh and the Royal Geographical Society, and to the Editors and Publishers of *Nature*, *Good Words*, and *Macmillan's Magazine*, for their freely accorded permission to reprint them. Most of the papers being records of geological rambles, I have introduced into them a few illustrative sketches from my notebooks.

EDINBURGH, *15th March 1882.*

CONTENTS.

		PAGE
I.	My First Geological Excursion	1
II.	"The Old Man of Hoy"	22
III.	The Baron's Stone of Killochan	40
IV.	The Colliers of Carrick	59
V.	Among the Volcanoes of Central France	74
VI.	The Old Glaciers of Norway and Scotland	109
VII.	A Fragment of Primeval Europe	145
VIII.	Rock-Weathering Measured by the Decay of Tombstones	159
IX.	In Wyoming	180
X.	The Geysers of the Yellowstone	206
XI.	The Lava-Fields of North-Western Europe	239
XII.	The Scottish School of Geology	250
XIII.	Geographical Evolution	272
XIV.	The Geological Influences which have affected the Course of British History	307

LIST OF ILLUSTRATIONS.

FIG.		PAGE
1.	"The Old Man of Hoy." (Sketched from the Sea)	26
2.	View of the Flagstone Cliffs of Holburn Head, Caithness	33
3.	View of Flagstone Cliffs, Brough of Birsay, Orkney	34
4.	View of the Gneiss Cliffs near Cape Wrath	37
5.	View of part of the Cliffs near St. Abb's Head	38
6.	View from the top of the Puy de Pariou	91
7.	Ice-worn bosses of gneiss and perched blocks. North coast of Sutherland	113
8.	Map of the Neighbourhood of the Holands Fjord (Munch)	117
9.	View of the two Glaciers of Fondalen, Holands Fjord	119
10.	Longitudinal Section of smaller Glacier. Fondalen	121
11.	Sketch-map of lower and of larger Glacier. Fondalen	123
12.	Sections across the lower end of the larger Glacier. Fondalen	124
13.	Map of the Jökuls Fjeld promontory (after Munch)	132
14.	View of Jökuls Fjord Glacier	134
15.	Section of Foot of Jökuls Fjord Glacier	136
16.	View of Glaciers at the head of Nus Fjord	141
17.	Section on beach at Nus Fjord	143
18.	Section on beach at Ardmarnock, Loch Fyne	143
19.	Ben Leagach, Glen Torridon	149
20.	View of the ancient platform of gneiss looking eastward from above Scourie, Sutherlandshire	151

LIST OF ILLUSTRATIONS.

FIG.		PAGE
21.	Ben Shieldag, Loch Torridon	153
22.	View of outlier of Cambrian breccia and sandstone among gneiss hills near Gairloch	155
23.	Sections of the junction of the fundamental gneiss and overlying Cambrian breccia. Gairloch	157
24.	Microscopic structure of white marble employed in Edinburgh tombstones	163
25.	Terraces of Great Salt Lake, along the flanks of the Wahsatch Mountains, south of Salt Lake City	211
26.	Alluvial Cones of the Madison Valley	215
27.	Terraces below the second cañon of the Yellowstone	219
28.	"Old Faithful" in eruption	229
29.	View on the Snake River, Idaho. Basalt Plain with younger volcanic cones	237

GEOLOGICAL SKETCHES

AT HOME AND ABROAD.

I.

MY FIRST GEOLOGICAL EXCURSION.[1]

'Tis an old story now, so far back, indeed, that I hardly like to reckon up the years that have since passed away. But clear and bright does it stand in my memory, notwithstanding, that quiet autumnal afternoon, with its long country ramble to an old quarry, the merry shouts of my schoolmates, the endless yarns we span by the way, and the priceless load of stones we bore homeward over those weary miles, when the sun had sunk, red and fiery, in the west, and the shadows of twilight began to deepen the gloom of the woods. Many a country ramble have I made since then, but none, perhaps, with so deep and hearty an enjoyment, for it opened up a new world, into which a fancy fresh from the Arabian Nights and Don Quixote could adventurously ride forth.

Up to that time my leisure hours, after school-lessons were learnt, and all customary games were played, had been given to laborious mechanical contrivances, based

[1] *Good Words*, 1861.

sometimes on most preposterous principles. For a while I believed I had discovered perpetual motion. Day and night the vision haunted me of a wheel turning, turning, in endless revolutions; and what was not this wheel to accomplish? It was to be the motive-power in every manufactory all through the country, to the end of time, to be called by my name, just as other pieces of mechanism bore the names of other inventive worthies, in that treasure of a book *The Century of Inventions*. Among various contrivances I remember striving hard to construct a boat that should go through the water by means of paddles, to be worked by a couple of men, or, failing them, by a horse; but though I found (if my memory serve me) that my hero, the old Marquis of Worcester, had anticipated the invention by almost 200 years, I could not succeed in getting the paddles to move except when the boat was out of the water, and so the grand contrivance, that might have made its discoverer famous in every harbour in the kingdom, fell to the ground.

The Saturday afternoons were always observed by us as a consecrated holiday-time, all school-work being then consigned to a delightful oblivion. To learn a lesson during these hours was regarded as something degenerate and wholly unworthy of the dignity of a schoolboy. Besides, we had always plenty of work of some kind to fill up the time, and what the nature of that work was to be for the ensuing Saturday had usually been determined long before the coveted Saturday came. Sometimes, if the weather was dull, my comrades repaired to my room (which we dignified as "the workshop") to hear a disquisition on the 'last invention, or to help if they could in removing some troublesome and apparently insuperable mechanical difficulty. Or we planned a glorious game of cricket, or golf, or football, that seldom came to a close until the evening grew too dark for longer play. In spring-time we would sally forth into the country to some well-remembered bank, where the primroses and violets bloomed earliest, and return at dusk, bringing many a bunch for those at home. The summer afternoons often found us loitering, rod in hand, along the margin of a shady streamlet, in whose deeper pools the silvery troutlet loved to feed. And it fed, truly, with little danger from us. The writhing worm (we never soared to the use of the fly), though ever so skilfully and unfeelingly twined round the hook, failed to allure the scaly brood, which we could see darting up and down the current without so much as a nibble at our tempting bait. Not so, however, with another member of that tribe, the little stickleback, or "beardie," as we called it, to which we had the most determined and unreasonable antipathy. The cry of "*A beardie! a beardie!*" from one of our party was the sign for every rod and stick to be thrown down on the bank, and a general rush to the spot where the enemy of the trout had been seen. Off went stockings and shoes, and in plunged the wearer, straight to the large stone in midchannel under which the foe was supposed to be lurking. Cautiously were the fingers passed into the crevices and round the base of the stone, and the little victim, fairly caught at last in his den, was thrown in triumph to the bank, where many a stone was at hand to end his torments and his life.

Autumn brought round the cornfields, and the hedgerows rich in hip, and haw, and bramble; and then, dear to the heart of schoolboy, came winter with its sliding, skating, and snowballing, and its long, merry evenings, with their rounds of festivity and plumcake.

'Tis an old story, truly; but I remember as if it had been yesterday, how my Saturday employments were changed, and how the vagrant, careless fancies of the schoolboy passed into the settled purposes that have moulded the man. I had passed a Saturday afternoon alone, and next day as usual met my comrades at church. On comparing notes, I found that the previous afternoon they had set out for some lime-quarries, about four miles off, and had returned laden with wonders—plants of strange form, with scales, teeth, and bones of uncouth fishes, all embedded in the heart of the stone, and drawn out of a subterranean territory of almost fabulous extent and gloom. Could anything more marvellous have been suggested to a youthful fancy? The caverns of the Genii, even that of the Wonderful Lamp, seemed not more to be coveted. At least the new cave had this great advantage over the old ones, that I was sure it was really true; a faint suspicion having begun to arise that, possibly, after all, the Eastern caverns might have no more tangible existence than on the pages of the story-book. But here, only four miles from my own door, was a real cavern, mysterious beyond the power of my friends to describe, inhabited by living men who toiled like gnomes, with murky faces and little lamps on their foreheads, driving waggons, and blasting open the rock in vast and seemingly impenetrable galleries, where the sullen reverberations boomed as it were for miles among endless gigantic pillars and sheets of Stygian water that stretched away deep and dark into fathomless gloom. And in that rock, wrapped up in its substance like mummies in their cerements, lay heaps of plants of wondrous kinds; some resembled those of our woods and streams, but there were many, the like to which my companions declared that even in our longest rambles they had never seen on bank, or brake, or hill; — fishes, too, there were, with strong massive scales, very different from our trouts and minnows. Some of the spiny fins, indeed, just a little resembled our foe the "beardie." Very likely (thought I), the Genius of the cave being a sensible fellow, has resolved to preserve his trout, and so with a murrain on the beardies has buried them bodily in the rock.

But above all, in these dark subterranean recesses lurked the remains of gigantic reptiles; and one of the quarrymen possessed a terrific tusk and some fragmentary scales, which he would have sold to my friends could their joint purse have supplied the stipulated price.

My interest in the tale, of course, increased at every new incident; but when they came to talk of reptiles, the exuberant fancy could contain itself no longer. "Dragons! dragons!" I shouted, and rubbed my hands in an ecstasy of delight. "Dragons, boys, be sure they are, that have been turned into stone by the magic of some old necromancer."

They had found too, in great abundance, what they had been told were "coprolites"— that is, as we afterwards learnt, the petrified excrement of ancient fishes. "*Copperlites*," thought I, nay, perchance it might be *gold;* for who ever read of such a famous cavern with petrified forests, fishes, and dragons, that had not besides huge treasures of yellow gold?

So there and then we planned an excursion for the following Saturday. The days that intervened stretched themselves somehow to an interminable length. It seemed the longest week of my life, even though every sleeping and waking hour was crowded with visions of the wondrous cavern. At length the long expected morning dawned, and soon brightened up into a clear, calm autumnal day.

We started off about noon; a goodly band of some eight or nine striplings, with two or three hammers, and a few pence amongst us, and no need to be home before dusk. An October sun shone merrily out upon us; the fields, bared of their sheaves, had begun to be again laid under the plough, and long lines of rich brown loam alternated with bands of yellow stubble, up and down which toiled many a team of steaming horses. The neighbouring woods, gorgeous in their tints of green, gold, and russet, sent forth clouds of rooks, whose noisy jangle, borne onward by the breeze, and mingling with the drone of the bee and the carol of the lark, grew mellow in the distance, as the cadence of a far-off hymn. We were too young to analyse the landscape, but not too young to find in every feature of it the intensest enjoyment. Moreover, our path lay through a district rich in historic associations. Watch-peels, castles, and towers looked out upon us as we walked, each with its traditional tales, the recital of which formed one of our chief delights. Or if a castle lacked its story, our invention easily supplied the defect. And thus every part of the way came to be memorable in our eyes for some thrilling event real or imaginary—battles, stern and bloody, fierce encounters in single combat, strange weird doings of antique wizards, and marvellous achievements of steel-clad knights, who rambled restlessly through the world to deliver imprisoned maidens.

Thus beguiled, the four miles seemed to shrink into one, and we arrived at length at the quarries. They had been opened, I found, along the slope of a gentle declivity. At the north end stood the kilns where the lime was burnt, the white smoke from which we used to see some miles away. About a quarter of a mile to the south lay the workings where my comrades had seen the subterranean

MY FIRST GEOLOGICAL EXCURSION.

men; and there too stood the engine that drew up the waggons and pumped out the water. Between the engine and the kilns the hillside had all been mined and exhausted; the quarrymen having gradually excavated their way southwards to where we saw the smoking chimney of the enginehouse. We made for a point midway in the excavations; and great indeed was our delight, on climbing a long bank of grass-grown rubbish, to see below us a green hollow, and beyond it a wall of rock, in the centre of which yawned a dark cavern, plunging away into the hill far from the light of day. My companions rushed down the slope with a shout of triumph. For myself, I lingered a moment on the top. With just a tinge of sadness in the thought, I felt that though striking and picturesque beyond anything of the kind I had ever seen, this cavern was after all only a piece of human handiwork. The heaps of rubbish around me, with the smoking kilns at the one end and the clanking engine at the other, had no connection with beings of another world, but told only too plainly of ingenious, indefatigable man. The spell was broken at once and for ever, and as it fell to pieces, I darted down the slope and rejoined my comrades.

They had already entered the cave, which was certainly vast and gloomy enough for whole legions of gnomes. The roof, steep as that of a house, sloped rapidly into the hillside beneath a murky sheet of water, and was supported by pillars of wide girth, some of which had a third of their height, or more, concealed by the lake, so that the cavern, with its inclined roof and pillars, half sunk in the water, looked as though it had been rent and submerged by some old earthquake. Not a vestige of vegetation could we see save, near the entrance, some dwarfed scolopendriums and pale patches of moss. Not an insect, nor indeed any living thing seemed ever to venture down into this dreary den. Away it stretched to the right hand and the left, in long withdrawing vistas of gloom, broken, as we could faintly see, by the light which, entering from other openings along the hillside, fell here and there on some hoary pillar, and finally vanished into the shade.

It is needless to recall what achievements we performed; how, with true boyish hardihood, we essayed to climb the pillars, or crept along the ledges of rock that overhung the murky water, to let a ponderous stone fall plump into the depths, and mark how long the bubbles continued to rise gurgling to the surface, and how long the reverberations of the plunge came floating back to us from the far-off recesses of the cave. Enough, that, having satisfied our souls with the wonders below ground, we set out to explore those above.

"But where are the petrified forests and fishes?" cried one of the party. "Here!" "Here!" was shouted in reply from the top of the bank by two of the ringleaders on the previous Saturday. We made for the heap of broken stones whence the voices had come, and there, truly, on every block and every fragment the fossils met our eye, sometimes so thickly grouped together that we could barely see the stone on which they lay. I bent over the mound, and the first fragment that turned up (my first-found fossil) was one that excited the deepest interest. The commander-in-chief of the first excursion, who was regarded (perhaps as much from his bodily stature as for any other reason) an authority on these questions, pronounced my treasure-trove to be, unmistakably and unequivocally, a fish. True, it seemed to lack head and tail and fins; the liveliest fancy amongst us hesitated as to which were the scales; and in after years I learned that it was really a vegetable—the seed-cone or catkin of a large extinct kind of club-moss; but, in the meantime, Tom had declared it to be a fish, and a fish it must assuredly be.

The halo that broke forth from the Wizard's tomb when William of Deloraine and the Monk of St. Mary's heaved at midnight the ponderous stone was surely not brighter, certainly not so benign in its results, as the light that now seemed to stream into my whole being, as I disinterred from their stony folds these wondrous relics. Like other schoolboys, I had, of course, had my lessons on geology in the usual meagre, cut-and-dry form in which physical science was then taught in our schools. I could repeat a "Table of Formations," and remembered the pictures of some uncouth monsters on the pages of our text-books— one with goggle-eyes, no neck, and a preposterous tail; another with an unwieldy body, and no tail at all, for which latter defect I had endeavoured to compensate by inserting a long pipe into his mouth, receiving from our master (Ironsides, we called him) a hearty rap across the knuckles, as a recompense for my attention to the creature's comfort. But the notion that these pictures were the representations of actual, though now extinct monsters, that the matter-of-fact details of our text-books really symbolised living truths, and were not invented solely to distract the brains and endanger the palms of schoolboys; nay, that the statements which seemed so dry and unintelligible in print were such as could be actually verified by our own eyes in nature, that beneath and beyond the present creation, in the glories of which we revelled, there lay around us the memorials of other creations not less glorious, and infinitely older, and thus that more, immensely more, than our books or our teachers taught us could be learnt by looking at nature for ourselves—all this was strange to me. It came now for the first time like a new revelation, one that has gladdened my life ever since.

We worked on industriously at the rubbish heap, and found an untold sum of wonders. The human mind in its earlier stages dwells on resemblances, rather than on differences. We identified what we found in the stones with that to which it most nearly approached in existing nature, and though many an organism turned up to which we could think of no analogue, we took no trouble to discriminate wherein it differed from others. Hence, to our imagination, the plants, insects, shells, and fishes of our rambles met us again in the rock. There was little that some one of the party could not explain, and thus our limestone became a more extraordinary conglomeration of organic remains, I will venture to say, than ever perturbed the brain of a geologist. It did not occur at the time to any of us to inquire why a perch came to be embalmed among ivy and rose leaves; why a sea-shore whelk lay entwined in the arms of a butterfly; or why a beetle should seem to have been doing his utmost to dance a pirouette round the tooth of a fish. These questions came all to be asked afterwards, and then I saw how egregiously erroneous had been our boyish identifications. But, in the meantime, knowing little of the subject, I believed everything, and with implicit faith piled up dragon-flies, ferns, fishes, beetle-cases, violets, sea-weeds, and shells.

The shadows of twilight had begun to fall while we still bent eagerly over the stones. The sun, with a fiery glare, had sunk behind the distant hills, and the long lines of ruddy light that mottled the sky as he went down had crept slowly after him, and left the clouds to come trooping up from the east, cold, lifeless, and gray. The chill of evening now began to fall over everything, save the spirits of the treasure-seekers. And yet they too in the end succumbed. The ring of the hammer became less frequent, and the shout that announced the discovery of each fresh marvel seldomer broke the stillness of the scene. And, as the moanings of the night-wind swept across the fields, and rustled fitfully among the withered weeds of the quarry, it was wisely resolved that we should all go home.

Then came the packing up. Each had amassed a pile of specimens, well-nigh as large as himself, and it was of course impossible to carry everything away. A rapid selection had therefore to be made. And oh! with how much reluctance were we compelled to relinquish many of the stones, the discovery whereof had made the opposite cavern ring again with our jubilee. Not one of us had had the foresight to provide himself with a bag, so we stowed away the treasures in our pockets. Surely practical geometry offers not a more perplexing problem than to gauge the capacity of these parts of a schoolboy's dress. So we loaded ourselves to the full, and marched along with the fossils crowded into every available corner.

Despite our loads, we left the quarry in high glee. Arranging ourselves instinctively into a concave phalanx, with the speaker in the centre, we resumed a tale of thrilling interest, that had come to its most tragic part just as we arrived at the quarry several hours before. It lasted all the way back, beguiling the tedium, darkness, and chill of the four miles that lay between the limeworks and our homes; and the final consummation of the story was artfully reached just as we came to the door of the first of the party who had to wish us good-night.

Such was my first geological excursion—a simple event enough, and yet the turning-point in a life. Thenceforward the rocks and their fossil treasures formed the chief subject of my every-day thoughts. That day stamped my fate, and I became a geologist.

And yet, I had carried home with me a strange medley of errors and misconceptions. Nearly every fossil we found was incorrectly named. We believed that we had discovered in the rock organisms which had really never been found fossil by living man. So far, therefore, the whole lesson had to be unlearned, and a hard process the unlearning proved to be. But (what was of infinitely more consequence at the time than the correct names, or even the true nature of the fossils) I had now seen fossils with my own eyes, and struck them out of the rock with my own hand. The meaning of the lessons we had been taught at school began to glimmer upon me; the dry bones of our books were touched into life; the idea of creations anterior to man seemed clear as a revealed truth; the fishes and plants of the lime-quarry must have lived and died, but when and how? was it possible for *me* to discover?

These quarries proved to our schoolboy band a never-ending source of delight. They formed the goal of many a Saturday ramble. The fishing-rod and basket gave place to hammer and bag; even our bats and balls and "shinties" were not unfrequently forsaken. Our love of legends, too, went on increasing, every walk giving rise to two or three new ones, extemporised for the occasion, and of course forgotten nearly as soon as invented.

Frequent visits made us better acquainted, not only with the quarries but with the quarrymen, and our ideas of the one were considerably influenced by our impression of the other. There were, I remember, three very distinct groups of workmen. The kilns at the north end were tended by a marked set of men. They seemed to be mostly Irishmen, whose duty it was to unload the waggons of limestone into the kilns, and keep up the supply of coal. The deep pits in which the rock was calcined sent up an intolerable heat, and gave out a thick, white, stifling smoke, that curled and drifted about with every veering of the wind. Creeping cautiously to within a short way of the edge of these fiery abysses, we could mark the red-hot rock cracking, and the coal flaming up from below it. The Irishmen, however, would march round the brink without a trace of fear or hesitation, and then, after the firing of the kilns, would squat themselves in the lee of a wall, an uncouth, sooty-faced company, each with a pipe, or else an oath, in his mouth. We never cultivated very closely the acquaintance of the kiln-men, an uneasy apprehension constantly arising that, on the slightest provocation, one of us might be tumbled into the pit, and never more be seen or heard of.

Very different in the nature of their work, and equally different in their disposition, were the men who tended the waggons which the engine drew up from the quarry. They had once worked below ground, but had now an easier post, their sole duty being to wheel off the full waggons as these came up, and to put empty ones on the rails to be let down the slope into the mouth of the excavation. One of them had lost a leg in his subterranean service, and was therefore somewhat slow in his movements. He had built himself a rude hut, with a fireplace and a wooden bench: and there I have often sat with him, and listened to his elucidation of the fossils, and his ideas of cosmogony in general. He was never at a loss for an explanation of any of the numerous fossils which he picked out of the limestone blocks that came up from the quarry. Some of his fellow-workmen maintained that rock and fossil were all created together, but my friend was a long way ahead of them. He was certain that the plants in the rock must have once bloomed green on the land, and that the fishes must have darted through the water. His Bible told him of a great flood that had destroyed mankind and covered the lands which they inhabited; and he had no manner of doubt that the fishes and plants of the limestone were memorials of that great inundation, and therefore contemporaries of Noah and the Ark.

The third, and by much the most numerous, group of workmen, were those whose labour went on underground—blasting and quarrying the limestone, and then wheeling it in waggons along the galleries to the mouth of the quarry, whence it was drawn up by the engine. Murky and grim, each with a slouched cap, from the front of which hung a little lamp, they formed, nevertheless, a merry company, keeping up a ceaseless din of hammering in these gloomy regions, save at intervals when a blast-hole was charged with gunpowder, and then all hurried away behind some of the huge pillars until the explosion was over. It was during one of these pauses that I first made their acquaintance. With one or two companions, I had been prying into the mouth of the quarry, and venturing for some way within, until, as the daylight grew dim, our courage failed, and we returned. A rumbling noise gradually approached, and there at last emerged from the darkness a full waggon, with a grimy workman pushing it from behind. The lamp that flickered on his forehead added greatly to his uncouthness as he came into the full light of day; and it was not without some hesitation that we accepted his invitation to hold on by the end of an empty truck, and return with him into the innermost recesses of the quarry. It was a long journey, and of course, save for the feeble glimmer of the lamp in his cap, in total darkness

Eventually we began to hear the sound of clinking hammers, and then in the dim distance we saw little lights moving to and fro. The sounds ceased as we approached, and the lights drew nearer, until we found ourselves in the centre of a group of begrimed workmen, which increased in numbers every moment as the men hurried from different parts of the workings to be out of the way of an impending blast.

They gathered round us, and examined our hammers as well as the specimens we had procured. One fossil had especially puzzled us, which we now submitted to the decision of our subterranean acquaintances. One of them —styled by his comrades "the Philosopher," a tall, wiry, young man—took the stone, and after eyeing it gravely for a few seconds, pronounced it to be an oyster-shell. I could see no resemblance on which to found such a decision; but the dictum of "Lang Willie" seemed to settle the matter finally in the eyes of the quarrymen. Seating himself on a large prostrate block of limestone, and stuffing his short pipe into his pocket, he proceeded to point out to the company the evidence that the scene of their labours had once been under the sea. There was the oyster-shell to begin with. Surely none of us could dispute that oysters only lived in the sea, and therefore, as the oyster occurred in the quarry, the quarry must once have formed part of the sea-bottom? Then there were the scales, bones, and teeth of fishes, very much longer than trout or any "siclike" fresh-water fish, and these must have dwelt in the sea. Besides this, he sometimes noticed a white powder crusting the rock like a sprinkling of salt, and the stone had occasionally what Trinculo would have called "an ancient and fishlike smell," that to Willie's mind clearly bespoke the former presence of the sea. All this and more was told at considerable length, with many a flourish of the fist, to the great apparent comfort and satisfaction of his brother-workmen.

And there was some truth in the reasoning. His facts, indeed, would not stand a very close scrutiny; even the little experience I had at the time enabled me to see their erroneousness; but his deductions, had the premises been sound, were fair enough. They showed, at least, a habit of thoughtfulness and observation much rarer among this class of men than we should expect to find it.

Such were my earliest clinical instructors in geology. With the help of their crude notions, added to our own boyish fancies, those of our number who cared to think out the subject at all strove to solve the problems that the quarry suggested. I cannot recall the process of inquiry among my comrades. But I well remember how it went on with myself. Our early identifications of all that we saw in the rock with something we had seen in living nature were unconsciously abandoned. I gradually came to learn the true character of most of the fossils, and recognised, too, that there was much which I did not understand, but might fairly attempt to discover. The first love of rarities and curiosities passed away, and in its place there sprang up a settled belief that in these gray rocks there lay a hidden story, if one could only get at the key.

There was no one within our circle of acquaintance from whom any practical instruction in the subject could be obtained. Probably this was a piece of good fortune for those of us who had the courage to persevere in the quest for knowledge. I can remember the long communings we had as to the nature of this or that organism, and its bearing on the history of the limestone. The text-books were of little service. So, thrown back upon ourselves, we allowed our fancy to supply what we could obtain in no other way. The ferns and other land-plants found in the limestone, together with the minute cyprids, of which the rock seemed in some places almost wholly composed, and the scales, bones, and teeth of ganoid fishes, indicated, as far as we could learn, that the deposit had accumulated in fresh water, perhaps in a lake or in the estuary of a river. But of course it was natural that we should try to discover what might have been the general aspect of the country when the animals and plants of the limestone were alive. We asked ourselves if the same hills existed then as now; if perchance the old river that swept over the site of the quarry took its rise among yonder pastoral glens; if the same sea rolled in the distance then as now, curling white along the same green shore. Happily ignorant of how far we had here ventured beyond our depth, it was not until after much questioning and disappointment that I found these problems to require years of patient research. The whole country for many miles round had yet to be explored, and minute observations to be made before even an approximation to a reliable answer could be given. But a boy's fancy is an admirable substitute for the want of facts. I did feel at times a little sorry that no evidence turned up on which to ground my restoration of the ancient topography of the district, or rather that such a world of work seemed to rise before me ere I could obtain the evidence that was needed. But the feeling did not last long. And so I conjured up the most glorious pictures of an ancient world, where, as in the land of the lotus-eaters, it was always afternoon, and one could dream away life among isles clothed with ferns and huge club-mosses, and washed by lakes and rivers that lay with-

c

out a ripple, save now and then when some glittering monster leapt out into the sunlight, and fell back again with a sullen plunge.

Happy afternoons were these! To steal away alone among the cornfields, and feast the eye on hill and valley, with their green slopes and bosky woods and gray feudal towers, and on the distant sea with the white sails speckled over its broad expanse of blue. And then when every part of that well-loved scene had been taken in, to let loose the fancy and allow the landscape to fade like a dissolving view until every feature had fled, and there arose again the old vanished lakes, and rivers, and palmy isles.

About two miles from the spot where we began our geological labours lay another quarry, from which lime had been extracted. When we first heard of it from our one-legged friend at the engine-house, we set it down as a continuation of his limework, the caverns of which seemed to run on underground to an indefinite length. There seemed nothing unlikely in the identification of two limestones only two miles distant from each other as part of one seam. So a Saturday afternoon was spent in the investigation of this second quarry.

Like the first, it had been opened along the slope of a gentle hill, and the excavations presented to our view a long line of caverns similar to those we had seen before. But the quarry was disused, and appeared to have been so for many years. The roof had fallen down in many places, the mouths of the caves had become well-nigh choked up with rubbish and tangled gorse, and the heaps of *débris*, so fresh and clean in our own quarry, were here overgrown with gray lichens and green moss, damp and old. The kilns had not been fired for many a day. The cracks and rents that had fissured their walls, from the fierce heat that once blazed within, were yawning hideously, as if a strong gale would hurl them with a crash into the half-buried cavern below. Only one human habitation was near, a small moss-grown cottage, where lived a little old woman, her skin brown and shrivelled as parchment, who was busy hanging out linen on a neighbouring hedge. Altogether, therefore, this second quarry had a very grave-like, antique look, and we entered it with a kind of boyish wonder whether so different a scene would yield us the same treasures as we had found so abundantly only two miles off.

It required but a cursory glance to show us that the two limestones were not the same. They differed in colour and texture, but still more in their fossil contents. We searched long but unsuccessfully for traces of the plants, or cyprids, or fish, so common at our first quarry. In their stead we hammered out an abundant series of quite different fossils, all quite new to us. Of course, in our attempts to discover the nature and habitats of these objects, we wandered quite as far from the truth as we had done before. After much blundering we eventually ascertained that the new treasures included corals, stone-lilies, and shells—all organisms of the sea-floor. But our most instructive collection of these relics of marine life were obtained from a much larger quarry some twelve miles away. This more distant locality was calculated to impress powerfully a much more matured imagination than that of boyhood. I have often since visited it, and always with fresh interest. It has quiet, tree-shaded nooks, where, the din of the workmen being hushed by distance, one may sit alone and undisturbed for hours, gathering up from the grass-grown mounds delicate lamp-shells and sea-mats, crinoids, cup-corals, and many other denizens of the palæozoic ocean. A mass of rock, from which the rest has been quarried away, stands in a secluded coppice, overlooking the sea, as if to show how thick the seam was before the quarrymen began to remove it. This mass has been exposed to the weather for many a long year. Its steep sides are crowded with stone-lilies, corals, and shells, which stand out in relief like an arabesque fretwork. The marks of the quarrymen's tools have passed away, and a gray hue of age has spread over the rock, aided by patches of lichen and moss, or by tufts of fern, that here and there have found a nestling-place. For here, as always, where man has scarped and wounded the surface of the globe on which he dwells,

> " Nature, softening and concealing,
> Is busy with a hand of healing."

From this point, between the overhanging branches, our schoolboy band could watch the lights and shadows flitting athwart the distant hills, the breeze sweeping the neighbouring sea into fitful sheets of darker blue, and the sails for ever passing to and fro. And then, turning round, there rose behind us this strange wall of rock—the bottom of an older sea, with its dead organisms piled by thousands over each other. I can never forget the impression made on my boyish mind by the realisation of this tremendous contrast in scenery and life, and of the vast gulf of time between the living world and the dead. It made a kind of epoch in one's life. My first afternoon in this old lime-quarry was of more service at this time than any number of books or lectures.

The recollection of these early days has often since impressed me with a sense of the enormous advantage which a boy or girl may derive from any pursuit that stimulates the imagination. My boyish geology was absurdly, grotesquely erroneous. I should have failed ignominiously at an examination which would be thought easy enough at a modern elementary science class. But I had gained for myself what these science classes so seldom infuse into the pupils—an enthusiastic love of the subject, and a determination to get somehow at the living truth of which the rocks are the records. I had learnt to treat fossils not as mere dead mineral matter, or as mere curiosities valuable in proportion to their rarity or perfection of preservation, but as enduring records of former life; not as species to fill a place in a zoological system, or specimens to take up so much room in a museum, but as the remains of once living organisms, which formed part of a creation as real as that in which we ourselves pass our existence. They were witnesses of early ages in our planet's history, and were ready to tell their tale if one could only learn how to read it from them. Few occupations possess greater power of fascination than to marshal all these witnesses, and elicit from them the evidence which allows us to restore one after another the successive conditions through which the solid land has passed. To realise how this is done, and to take part in the doing of it, is for a boy a lifelong advantage. He may never become a geologist in any sense, but he gains such an enlarged view of nature, and such a vivid conception of the long evolution through which the present condition of things has been reached, as can be mastered in no other way. A single excursion under sympathetic and intelligent guidance to an instructive quarry, river ravine, or sea-shore, is worth many books and a long course of systematic lectures.

II.

"THE OLD MAN OF HOY."[1]

THE tidal wave of travellers, which, thanks to railroads and steamboats, pours northward over the country every summer, even as far as John o' Groat's, has as yet hardly risen much beyond that utmost shore. The tourist stops short at the Pentland Firth; indeed, when he reaches its bare treeless coast, and finds that there is really no traditional house at John o' Groat's (though a good inn, with careful host and kindly hostess, should tempt him to rest there a while), he is in a hurry to get back by daylight to the busy hum of men in the hyperborean city of Wick or of Thurso, and as eager to flit southwards again next morning. He makes a fatal mistake, however; for he misses the very points which it would have been worth his while to make the whole of his long journey to see. Let him, for instance, take up his quarters for a day or two by the side of the Pentland Firth, and spend his hours watching from one of its grim cliffs the race of its tideway. Nowhere else round the British Islands can he look down on such a sea. It seems to rush and roar past him like a vast river, but with a flow some three times swifter than our most rapid rivers. Such a broad breast of rolling eddying foaming water! Even when there is no wind, the

[1] *Geological Magazine*, 1878.

tide ebbs and flows in this way, pouring now eastwards now westwards, as the tidal wave rises and falls. But if he should be lucky enough to come in for a gale of wind (and they are not unknown there in summer, as he will probably learn), let him by no means fail to take up his station on Duncansbay Head, or at the Point of Mey. The shelter of a flagstone "dyke" and a waterproof will save him from any ulterior consequences of the exposure, or should he have some misgivings on this point, he will find, when he gets back to the shelter of the inn at John o' Groat's, that mine host has sundry specifics of well-tried potency, at the very sight and taste of which rheums, catarrhs, and the rest of that tribe of ailments at once decamp. Ensconced in his "neuk," he can quietly try to fix in his mind a picture of what is before him. He will choose if he can a time when the tide is coming up against the wind. The water no longer looks like the eddying current of a mighty river. It rather resembles the surging of rocky rapids. Its surface is one vast sheet of foam and green yeasty waves. Every now and then a huge billow rears itself impatiently above the rest, tossing its sheets of spray in the face of the wind, which scatters them back into the boiling flood. Here and there, owing to the configuration of the bottom, this turmoil waxes so furious that a constant dance of towering breakers is kept up. Such are the terrible "Roost of Duncansbay," and the broken water grimly termed the "Merry Men of Mey." With a great gale from the north-east, or south-east, the shelter even of the stone wall on Duncansbay Head would be of little avail. For solid sheets of water rush up the face of the cliffs for more than a hundred feet, and pour over the top in such volume that it is said they have actually been intercepted on the landward side by a dam across a little valley, and have been used to turn a mill. Should the meditative tourist be overtaken by such a gale he will find shelter in the quaint cottage of the kind-hearted but hard-headed John Gibson, who, perched like a sea-eagle at the head of a tremendous chasm in the cliffs, can spin many a yarn about the tempests of the north.

No one can see such scenes without realising, as he probably has never done before, the restless energy of nature. His eyes are opened. He feels how wind and rain, wave and tide, are leagued together, as it were, in spite of their apparent antagonism, to batter down the shores. Everywhere he witnesses proofs of their prowess. Tall gaunt stacks rise out of the waves in front of the cliffs of which they once formed a part. Yawning rents run through them from summit to base; their sides are frayed into cusp and pinnacle that seem ready to topple over when the next storm assails them; their surf-beaten basements are pierced with caverns and tunnels into which the surge is for ever booming. On the solid cliffs behind, the same tale of warfare is inscribed. But the traveller who has seen so much will perforce desire to see more. From his perch on the southern side of the foaming Pentland Firth he looks across to the distant hills of Hoy—the only hills, indeed, which are visible from the monotonous moorlands of northern Caithness, save when from some higher eminence one catches the blue outline of Morven on the southern sky-line. The Orkney Islands are otherwise as tame and as flat as Caithness. But in Hoy they certainly make amends for their generally featureless surface. Yet even there it is not the interior, hilly though it be, but the western coast-cliffs, which redeem the whole of the far north of Scotland from the charge of failure in picturesque and impressive scenery. One looks across the Pentland

Firth and marks how the flat islands of the Orkney group rise from its northern side as a long low line, until westwards they mount into the rounded heights of Hoy, and how these again plunge in a range of precipices into the Atlantic. Yellow and red in hue, these marvellous cliffs gleam across the water as if the sunlight always bathed them. They brighten a gray day, and gray days are only too common in the northern summer; on a sunny forenoon, or still better on a clear evening, when the sun is sinking beneath the western waters, they glow and burn, yet behind such a dreamy sea-born haze, that the onlooker can hardly believe himself to be in the far north, but recalls perhaps memories of Capri and Sorrento, and the blue Mediterranean. Looking at them from the mainland, we are soon struck by one feature at their western end. A strange square tower-like projection rises behind the last and lowest spur of cliff which descends into the sea. We may walk mile after mile along the Caithness shore, and still that mysterious mass keeps its place. As we move westwards, however, the higher cliffs behind open out, and we can see on a clear day with the naked eye that the mass is a huge column of rock rising in advance of the cliff. It is the "Old Man of Hoy"—a notable landmark, well deserving its fame.

Let no tourist who has journeyed as far as Thurso hesitate to cross the Firth and reach Stromness in Orkney. He will find a steamer ready to carry him thither in a few hours, and in the voyage will pass close under the grandest cliff in the British Islands. Above all, he will make the personal acquaintance of the Old Man, or at least will be brought so near as to conceive a very profound respect for him. The view given in Fig. 1 was sketched from the vessel in this passage, and though by no means taken from

Fig. 1.—"THE OLD MAN OF HOY."
(Sketched from the sea.)

the most picturesque point of view, may serve to convey some notion of the form, size, and structure of this the most remarkable feature in Orkney scenery. The Old Man is a column of yellow and red sandstone more than 600 feet high. It stands well in front of the cliff, with which, however, it is still connected by a low ridge strewn with blocks. Doubtless one main cause of its impressiveness lies in the fact that its summit is considerably higher than the cliff behind it. Thus it stands out against the sky even when seen from a distance. Its base is washed on three sides by the waves which rise and fall over a low reef running out from underneath the base of the column. Formerly a huge buttress, like the Giant's Leg of Bressay in Shetland, used to project into the sea. But it has been swept away, and for many years the Old Man, with the support of but one leg, has had to keep his watch and wage his unequal battle with the elements.

Unless the ground-swell be too heavy, the steamboat usually keeps close enough to the base of the great precipices to allow the masonry of this wonderful obelisk to be distinctly seen. Like the cliff behind, it is built up of successive bars of sandstone forming portions of horizontal or very gently inclined strata. Its base, however, rests on a pedestal of different materials, consisting of two well-defined bands, both of which can be traced stretching landwards and passing under the base of the cliff. The lower of these two bands is plainly marked by lines of parallel stratification inclined at a considerably higher angle than the dip of the sandstones, and evidently composed of something quite different from them. Viewed thus from the sea in a brief and passing way, the whole structure can be recognised as composed of three distinct portions. The main pillar, of pale red and yellow sandstone, rests unconformably upon a platform composed of two layers, of which the uppermost is a dark band of seemingly structureless rock, while the lower is formed of dark slate-coloured tilted strata.

It is only when one lands on the island of Hoy, and examines the cliffs in detail, that the true nature and history of the three bars of the Old Man can be made out. The yellow and red sandstones of the column and the cliff behind it are then found to present the ordinary characters of the Upper Old Red Sandstone, to which they are with probability referred, though as yet they have yielded no fossils. Irregularly alternating in thick and thinner beds, they are rent by innumerable perpendicular joints. By means of these divisional lines, slice after slice falls away from the face of the cliffs, which thus maintain their precipitous front towards the Atlantic. Except in regard to their scenic features, these sandstones, however, are less full of interest than the two bars comprising the Old Man's pedestal. The upper bar consists of a band of dark amygdaloidal lava with a slaggy surface. The same rock appears elsewhere, rising out from beneath the sandstones of the precipices, particularly at the north-western headland, where it consists of three or more distinct bands with well-stratified volcanic tuffs. To the north-east of that headland, on a tract of lower ground intervening between the base of the hills and the edge of the sea, several well-marked volcanic "necks" or pipes occur, representing some of the vents from which the streams of lava and showers of ash were poured. The complete interstratification of the beds of erupted material with the lower portion of the sandstones proves that the volcanic action showed itself at the beginning of the deposition of the Upper Old Red Sandstone in this region. Another little vent may be observed on the Caithness coast, near John o' Groat's

House. Perhaps some may still remain to be noticed among the central and northern members of the Orkney Islands. It seems to have been a singular and local outburst of volcanic energy during Upper Old Red Sandstone times—the only one yet discovered to the north of the Highlands. The uppermost bar, then, of the pedestal on which the Old Man has taken his stand is a massive sheet of lava.

The lower bar belongs to a very different period, and has a totally dissimilar history. Its component strata have been upturned and worn away before the eruption of the lava, which had rolled over their broken and bared edges. On looking more closely into these strata, which, even seen from the sea, present such a contrast in disposition to the lava and overlying sandstones, we find that they consist of dark thin-bedded sandstones, shales, and impure limestones. In short, they are a portion of the great series of deposits known as the Caithness flagstones of the Lower Old Red Sandstone. From many of their exposed surfaces shining jet-black scales, bones, and teeth of the characteristic fishes of these flagstones project. What a suggestive picture of the imperfection of the geological record is presented to us by some of these weather-beaten or surf-worn sheets of rock! We pick up from their crannies broken whelks, nullipores, and corallines, tossed up by the last storm from the zones of life now tenanting the sea below us. The limpet and sea-anemone, the whelk and barnacle, are clinging to the hardened sand over which, while it was still soft, the *Osteolepis* and *Coccosteus* and their bone-cased brethren disported in the ancient northern lake of Lower Old Red Sandstone times. Nay, we may now and then watch a living mollusc creeping over the cuirass of a palæozoic fish. Yet who can realise the lapse of time which here separates the living from the dead?

Below and beyond the horizon of the flagstones no evidence among the Hoy cliffs remains to lead us. But in the neighbouring isles of Pomona and Gremsa, bosses of crystalline rocks — granite, gneiss, and schists — project from under the flagstones, and are wrapped round with conglomerates, doubtless representing islets with the shore-gravel heaped up around them when they rose out of the Old Red Sandstone lake.

So much for the materials out of which the Old Man has been carved. And now a few words as to the process of carving. If the traveller who has reached Stromness finds himself with even one spare day at his disposal, he cannot employ it to more conspicuous advantage than by taking a boat with a couple of stalwart Norse like Orcadian boatmen, crossing the strait to Hoy, and ascending that island by the Cam and the north-western headland, with its rock-girt corry and glacier-moraines, until he finds himself at the summit of the great western precipice, with the surface of the surging Atlantic some 1300 feet below him. The scene tells its own tale of ceaseless waste, and needs no lecture or text-book for its comprehension. Pinnacles and turrets of richly-tinted yellow and red sandstone roughen the upper edge of the cliff, often fretted into the strangest shapes, and worn into such perilous narrowness of base that they seemed doomed to go headlong down into the gulf below when the next tempest sweeps across from the west. Buttresses, sorely rifted and honey-combed, lean against the main cliff as if to prop it up; but separated from it by the yawning fissures which will surely widen until they wedge off the projecting masses, and strip huge slices from the face of the cliff. One sees, as it were, every step in the progress of degradation. It is by this prolonged splitting and slicing and fretting that the precipice has been made to recede, and has acquired its shattered but picturesque contours. The Old Man is thus a monument of the retreat and destruction of the cliffs of which it once formed a part. To what accidental circumstance it may have owed its isolation cannot be affirmed with certainty. But it shares in the prevalent decay. Every year must insensibly tell upon its features.

On the calmest day some motion of air always plays about the giddy crest of these precipices, and a surge with creaming lines of white foam sweeps around their base. But when a westerly gale sets in, the scene is said to be wholly indescribable. The cliffs are then enveloped in driving spray torn from the solid sheets of water which rush up the walls of rock for a hundred feet or more, and roll back in thousands of tumultuous waterfalls. The force of the wind is such as actually to loosen the weathered parts of the rock and dislodge them. Thus along the mossy surface of the slope, which ascends inland from the edge of the cliff, large flat pieces of naked stone may be picked up by scores lying on the heather and coarse grass, whither they have been whirled up from the shattered crags by successive gusts of the storms.

The destruction of this coast-line has not yet, however, wholly effaced traces of other powers of waste which have long since passed away. On the very edge of the cliff, to the south-east of the Old Man, some well-preserved striations on the sandstone point to the movement of the ice-sheet of the glacial period across even the hilly island of Hoy in a N.W. and S.E. direction. Again, in the green corry at the Cam of Hoy, some beautifully perfect little moraines remain to show that after the great land-ice had subsided the snow-fall in these northern regions continued heavy enough to nourish in so small an island as Hoy groups of valley glaciers. Though the general form of the hills and valleys remains now much as it was when the last lingering glacier melted away, there have been stupendous changes since then in the shaping of the precipices. At that time the Old Man still formed a portion of the solid cliff. It is in the ensuing interval that this impressive landmark has been left during the destruction of the surrounding masses. Long may he be able to stand his ground! When his last hour comes, as come it must, may some reverential geologist, duly impressed with a sense of the might of denudation in the sculpture of the land, be there to pay the last honours to his dust!

In the scenery of the British Islands no geological formation plays a more varied part than the Old Red Sandstone, and nowhere can its characteristic landscapes be more instructively seen than in these far northern districts of Scotland. In Hoy, for example, the upper sandstones rise into a group of smooth dome-shaped hills, which, from all sides, stand out in striking contrast to every other form of ground within sight. In Caithness the lower sandstones and conglomerates have concentrated all their efforts on the production of the one solitary mountain of that county —Morven—a graceful cone, which so towers above the moors on the one side and the sea on the other as to form one of the most notable landmarks in the north of Scotland. But on the coast-line, where the rocks assert most strongly their individuality of character, swept bare of all protecting soil by the restless and resistless surge, their minutest points of structure are so exposed as to affect even the most delicate lineaments of the cliffs. The two fundamental structures, bedding and jointing, are developed with a trenchant emphasis which gives a dominant character to the scenery of the shores of Caithness and Orkney. Walls of flagstone, several hundred feet high, are seen from base to summit to consist of thin parallel bands of horizontal or gently-inclined strata. These beds, though everywhere singularly durable, vary slightly in their powers of resistance to the elements. The less tenacious layers are eaten away, while the harder project beyond them. Hence the precipices

Fig. 2.—View of the Flagstone Cliffs of Holburn Head, Caithness, showing how their vertical face is produced by the wedging off of successive slices of rock along lines of joint.

are fretted into alternate lines of cornice and frieze, which can be followed by the eye from buttress to buttress along the front of these grim cliffs.

That the flagstone must, however, be endowed on the whole with exceptional durability is shown by the striking verticality which the precipices maintain. Their perpendicular walls are defined by the system of joints which always traverse the rock vertically or at high angles. Slice after slice is wedged off by means of these joints, and in this way the perpendicular front of the cliffs is maintained. In many places the observer may watch the process of sculpture in successive stages of progress. He will notice that,

Fig. 3.—View of Flagstone Cliffs, Brough of Birsay, Orkney, showing how the overhanging form of a precipice may be determined by the inclination of rock-joints.

as a rule, the dismemberment begins at the top of the cliff, where the agents are not the breakers of the ocean, but rain, frost, and the other powers of the air. A joint may be observed to gape a little at the summit of the precipice, where nature's wedge has begun to be driven home. In another case the wedge has gone down to the very base of the cliff. The disjointed buttress is severed from the main mass by a yawning rent, which will be slowly widened above, while the breakers breach it below, until the whole will fall into the surf, and expose the naked cliff behind to a repetition of the same waste.

If the joints are vertical the resulting face of precipice will be vertical also (Fig. 2); and this fact, combined with the singular durability of the flagstone, accounts for the sheer walls by which so much of Caithness and Orkney is girdled round. Any deviation from verticality in the joints will of course produce a corresponding departure in the resulting cliff. Hence where, as often happens in these regions, the joints are slightly inclined landwards, the precipices are actually made to overhang. In such cases it is easy to show that the beetling walls are not really eaten away faster by the waves below than by the subaerial agents above (Fig. 3).

Another singular feature of these northern coasts is the number of *gios*, or narrow steep-walled gullies, or inlets, by which the sea-cliffs are indented. Here again we trace the dominant influence of the joints. In fact, the waste of these shores may be compared to a gigantic process of quarrying, wherein the rains, snows, and frosts above, the springs and trickling water within, and the breakers below, are the unwearying workmen. Whether the sea-wall is demolished uniformly, or portions of it are allowed to remain as projecting buttresses, or isolated into massive quadrangular sea-stacks, or cut into deep narrow recesses, nature works along the joints as quarrymen would do, and thus the massive architectural character of these cliffs is preserved. At the same time the slow progress of atmospheric waste sculptures the bare wall of rock into its characteristically striped and fretted surface, and brings out the peculiar weather-tint of each bed, from deepest indigo to palest emerald-green. On some of the ledges a scanty vegetation finds root, and where the cliffs rise most inaccessibly from the waves each cornice along their front is the nestling-place of innumerable sea-birds, whose shrill screams blend with the sough of the wind and the monotonous cadence of the surge into a wild northern music that wakens many a chord in the heart of one to whom the elemental sounds of nature are ever dear. No sooner do we step off the Old Red Sandstone than these singularly characteristic and persistent features disappear. The contrast presented by some of the other rocks of the North must strike every observer, even one to whom the very name of geology is unknown. The traveller who journeys westward into Sutherlandshire encounters many varieties of coast scenery, but he leaves behind him the peculiar cliffs of the Caithness flagstones. At one point he is confronted with gleaming precipices and steep acclivities of white glistening quartzite, at another he beholds vast sea-walls of a sombre dull red sandstone, even more colossal than those of Caithness, but wanting in those charms of light and shade, wealth of colour, and multiplicity of detail in form, which give the flagstone scenery so defined a character. Perhaps the greatest contrast is to be seen among the gneiss precipices of Cape Wrath. That north-western headland of Scotland is composed of the oldest rock in Britain, and one that from its tough, massive, gnarled aspect is well worthy of its position as the foundation on which the geological structure of these islands has been erected. Rising into a range of singularly scarped and rugged cliffs, it bears the full brunt of every storm that sweeps across the open Atlantic. Every weak part of its framework is discovered by the powerful battery of breakers, and is hollowed

Fig. 4.—View of the Gneiss Cliffs near Cape Wrath, with tortuous granitic veins.

into tunnel, cave, or gully, while the harder parts tower up into fantastic columns or buttresses. It possesses no symmetry of structure like that of bedding and jointing among the flagstones. Huge tortuous veins of a coarse kind of granite run up the face of the cliffs, reminding one of the prominent sinews of some antique statue (Fig. 4). The main mass of the rock through which these veins interlace is of a dull dusky green or livid gray, while the veins them-

Fig. 5.—View of part of the Cliffs near St. Abb's Head, showing curved Silurian strata.

selves stand out in pale flesh colour, so that even from a distance of several miles this singular feature of the cliffs may be distinctly seen.

As still another illustration of the intimate dependence of our rocky coast scenery upon geological structure, reference may be made to the range of cliffs on the south-eastern margin of Scotland on either side of St. Abb's Head. Here the bedding of the rocks is almost as plainly marked as among the flagstones of Caithness. But the strata, instead of lying in horizontal or gently-inclined undisturbed succession, have been thrown into huge folds which sweep from summit to base of precipices sometimes 500 feet high (Fig. 5). The lines of stratification consequently curve to and fro among the cliffs, carrying with them their successive bars of massive graywacke or fissile shale. An intricate system of minor cross-joints causes these bands of rock to split up into irregular blocks, while by a set of large but somewhat ill-defined joints the cliffs are cleft into vast irregular bastions and recesses. On one of these projecting crags the ruined fortalice of Fast Castle—the prototype of Scott's Castle of Ravenswood—is perched. Here and there at the base of the cliffs are sheltered caves, once favourite haunts of smugglers, now hardly ever disturbed by human voices. Gaunt sea-stacks, once part of the main cliff, but now isolated amid the surf, stand up in front and are favourite resting-places for crowds of sea-fowl. On all these rock-faces, whether main precipice or detached outlier, the peculiar contour of the curved strata may be traced, giving the scenery a character of its own, which only reappears with the recurrence of the same kind of geological structure.

III.

THE BARON'S STONE OF KILLOCHAN.[1]

On a gentle green declivity that slopes down to the Water of Girvan, and within sight of the broad Firth of Clyde, which the Girvan enters only three miles farther down the valley, stands a large gray block of granite, known in the district as the Baron's Stone of Killochan. From this stone looking seaward, on a clear day, when a breeze from the north-west has freshened the Firth into deepest azure, you can see, far away beyond the bold headlands of Carrick, the long blue lines of the hills of Antrim. And if you go but a few yards up the hill you may trace these faint promontories vanishing into the west, and then the long low hills of Cantyre bounding the western horizon, while in the midst of the wide stretch of sea Ailsa Craig lifts its scarred sides 1100 feet above the surf that beats about their base. The nearer landscape is formed by the valley of the Girvan, narrow and straight, with a ridge of green hills about 1000 feet high on the south side, a range of lower wooded eminences on the north, and the river winding in endless curves along the bottom. Looking up this valley, the eye wanders with delight over a mingled grouping of woodland and meadow, revealing here and there a reach of the blue stream and a strip of soft bright pasture.

[1] *Macmillan's Magazine*, XVII., 1868.

The woods climb up boldly along the hillsides, overshadowing every little dingle and watercourse, and so sweeping onwards up the valley, in every tint of green, and every variety of mass and outline, until a bend of the hills closes in the view. Even as a piece of scenery, this vale of the Girvan, though less known than many others in the lowlands of Scotland, has a charm which these often want. There is one respect, at least, wherein it has a peculiar interest. I know of few Scottish landscapes so circumscribed in extent, yet into which are crowded so many human associations of bygone times. On the hill-tops that look down upon us are the mouldering ramparts of the earthen forts of the early races. From the lower grounds the plough and harrow have long effaced such antique memorials: but the traditions of the primitive people survive in the very names of the hamlets and meadows. From these names we learn of Culdee saints to whom shrines were erected all down the course of the Girvan. And we see how the natives were Celtic, speaking the same language that still survives in the Highlands, and displaying the same nice discrimination and poetic turn of thought in the choice of names for their rivers, and crags, and hills. The castles of feudal times have survived better in this district of Ayrshire than in most other parts of Scotland. There are the remains of at least a dozen of them in the lower sixteen miles of the Girvan valley. Most of these, indeed, are ruinous; but some still form part of more modern mansions, and at least one—the old house of Killochan—remains nearly as it was some three hundred years ago. Nor are these merely interesting from their antiquity. Each is linked more or less closely with the history of the district, and sometimes not of the district only but of the kingdom at large. For the barons of

Carrick were a warlike race, ever at feud either with each other or with their neighbours in the adjoining sheriffdoms, and they had power enough to make themselves of consequence for good or ill to the government of the realm. But of the barons more anon.

Looking at the great size and weight of the Stone of Killochan, one is tempted at the very first to ask how so large a block came to be where it now lies. It measures roughly about 480 cubic feet, and must thus weigh somewhere about thirty-seven tons. There are no overhanging crags from which it could have rolled. It stands high above the river, and fully 100 feet above the sea, so that we can scarcely imagine it to have been washed down by floods, even if its great size did not forbid such a supposition. But our surprise increases when we find that this great mass of rock consists entirely of a close-grained granite. There is in the neighbourhood no granite hill from which it could have been detached. Silurian grits, slates and limestones, Old red sandstones and conglomerates, Carboniferous shales, freestones and coals, form all the surrounding country; but there is no granite. Whence, then, came the Baron's Stone? Perhaps a casual visitor might be bold enough to imagine that it was brought up from the coast by some of the old barons, having been shipped across from Arran. The size of the boulder, however, is enough of itself to show the absurdity of such a notion. Let the visitor step down to the margin of the river and look at the blocks of granite—less, indeed, in size, but similar in composition and form—which are lying by scores along the watercourse. Let him turn eastward into the picturesque little dell, by the side of which lies the carriage-way to the castle. In the bed of the rivulet he will see another set of large granite boulders, one of them containing about 200 cubic feet of stone. Throughout the whole valley, in short, he can hardly turn anywhere without encountering similar boulders. They have been mostly cleared off the cultivated places, and may be seen gathered into groups at the corners of the fields. They crowd the bottom of all the streamlets. The field-fences are built of them; road walls, doorposts, lintels, even entire cottages, have been made out of these widely-distributed stones. The old barons would have had but a sorry time of it had their days been spent in bringing granite boulders from a distance to mar their own fields and cumber their moors and hillsides, already barren enough by nature. They could then have enjoyed but little leisure for the pastime of killing and maiming each other. And yet all the barons of Carrick, with all their vassals and retainers to boot, working hard together for five hundred years, could not have done a thousandth part of the work.

So conspicuous a feature in the scenery of the country could not well escape notice, especially in early times, when a supernatural origin was easily found for what could not otherwise be readily explained. I have not yet been able to recover any of these traditional theories about the boulders in this part of Scotland. They still exist, however, in other districts; and, as a good sample of the class, especially in the way of showing the dry humour which enters so largely into elfin legend north of the Tweed, I may quote one which came under my own notice some time ago in Clydesdale. Not many miles above the Falls of Clyde the river makes some serpentine curves through a wide alluvial plain. One of these bends approaches the village of Carnwath, and the stream has there cut away part of a bank of soft clay and sand, on which are scattered a number of blocks of greenstone. An intelligent native of Carnwath, to whom I applied for information about the former number of boulders, told me that in his boyhood the ground between the river and the Yelping Craig, about two miles off, was literally strewed over with blocks of all sizes, up to masses six feet or more in height. So abundant were they to the south-west of Carnwath, that one tract was known as the "Hell Stanes Gate," *i.e.* road, and another as the "Hell Stanes Loan." The stones have now well-nigh disappeared under the sway of the farmers, but the old legend of their origin still remains. My informant, after pointing out the graves of some of the larger boulders, and the broken remains of others, went on to tell how, in old times, Michael Scott and the devil had entered into a compact with a band of witches to dam back the Clyde. It was one of the conditions of the agreement that the name of the Supreme Being should never on any account be mentioned. All went well for a while; some of the more stalwart spirits having brought their burden of boulders to within a few yards from the edge of the river, when one of the younger members of the company, staggering under the weight of a huge block of greenstone, exclaimed, "O Lord, but I'm tired." Instantly every boulder tumbled to the ground, nor could either witch, warlock, or devil move a single stone one step thereafter. And there the blocks lay for many a long century, until the industrious farmers quarried and blasted and buried them.

There can be little doubt that the elfins of old were not less busy in Carrick, though the records of their doings have faded from tradition. It is still told, however, that one witch, of more than ordinary audacity and strength, lifted a great hill from the Ayrshire uplands, and, putting it in her apron, made off through the air for Ireland. But, as bad luck would have it, the apron-strings broke on the passage, and the hill fell with a fearful plunge into the Firth, where it still remains, under the name of Ailsa Craig. The only original account of the boulders of the Girvan valley which has come under my notice was that of a mason, who, when asked his idea of the endless blocks of granite that dot the fields and hillsides like flocks of sheep, gravely remarked that "when the Almichtie flang the warld out, He maun ha'e putten thae stanes upon her to keep her steady."

Supernatural agency failing us, we come back again to the question, Whence came the Baron's Stone of Killochan and all its kindred boulders? There is, as every tourist knows, a great mass of granite in Arran. It rises into the noble cone of Goatfell, and forms the chains of jagged peaks that overshadow the defiles of Glen Rosa and Glen Sannox. But this granite is not the same as that of the Carrick boulders. It differs in texture, partly also in composition, and in certain mineralogical peculiarities which need not be specified here. There can, indeed, be no doubt whatever that the boulders did not come from Arran. Where, then, is their source to be sought? Let us in imagination make our way up the valley of the Girvan, and note as we go such changes of scenery and rock as may chance to throw light on the matter. The lower or seaward portion of the river's course runs along the northern base of a tolerably steep line of hills, rising, as I have said, to heights of over a thousand feet, and sweeping away southward and eastward into the wild mountainous uplands of Carrick and Galloway. After skirting these hills for about sixteen miles, among woodlands and pleasure-grounds, and past the remains of ancient strongholds, the course of the stream bends round at nearly a right angle towards the south, and enters the uplands through a narrow and deep defile. Looking up this straitened valley the cultivated country lies all behind us, while in front are the lonely hills. The change of scenery takes place so suddenly that no sooner do we plunge into the chain of hills than the rich woods and cornfields disappear, steep grassy and rocky declivities descend abruptly upon the narrowed strip of alluvial soil that borders the river; trees occur only at intervals, and chiefly down the watercourses; the herbage grows more and more heathy, and traces of cultivation more and more scanty, until, as we wind up the valley, we at last take leave of all signs of human habitation, and enter upon a region of wide, desolate, treeless moorland, and gray craggy mountain. The lower parts of the course of the Girvan lie chiefly upon the various members of the Carboniferous series of rocks. But the upper portion, which winds through the high grounds, has been hollowed out of the northern margin of the wide band of Silurian rocks stretching entirely across the south of Scotland from the Irish Sea to the German Ocean. These Silurian strata, bent and broken like crumpled parchments, presenting at the surface every variety of crag and knoll, dingle and dell, rounded hill, steep precipice, and rough, rugged mountain, form the whole of the wide uplands of Carrick and Galloway, where they mount to a height of more than 2700 feet above the sea. It is on the northern flank of the highest chain of the great central group of hills that the Girvan has its source. Following its course upwards from the lowland country, we find the same abundance of boulders in the narrowed valley as in the more open parts towards the sea. Still we fail to trace any granite forming a solid part of a hill. Conglomerate, shale, grit, porphyry, and other kinds of rock, crop out along the sides of the glens, but without any symptoms of granite. And yet the granite boulders, gray and lichened, are strewed over these hillsides, just as they were seen far down over the Carboniferous strata of the low grounds. At a height of between 700 and 800 feet above the sea there are some remarkable mounds on our way, formed of loose earth and clay, with abundance of boulders of various Silurian rocks, and here and there with large blocks of granite strewed over their surface. Similar mounds occur higher up, and all the interval is studded as usual with granite boulders. Still we can see no granite in place. Passing one or two small lakes or lochans, which receive and discharge the waters of the Girvan in an undulating mossy tract of ground, we begin to be utterly amazed at the prodigious quantity as well as the great size of the granite blocks. Gray and lichen-crusted, or crumbling into sand, they are scattered over the valley by thousands. They lie on all manner of declivities, sometimes on mounds of rubbish, sometimes on prominent ridges of rocks, and sometimes half-buried in peat-bogs, like groups of "laired" cattle. Moreover, as we rise with this broken ground, our eyes are struck with the strange hummocky shapes into which the hillsides have been worn. The solid rock comes almost everywhere to the daylight in the form of rounded knolls and hollows, which, especially where they have been preserved from the wear and tear of the weather by a coating of turf or soil, have a singularly smoothed and polished appearance, which is rendered all the more marked, seeing that the edges of the vertical strata have been ground down into one common undulating surface. On such rounded and polished bosses of rock the never-failing granite boulders may be seen at every turn. At length the valley narrows in a scene of strange lonely grandeur. The brawling brook—it no longer merits the title of river—throws its amber waters into foam over endless boulders that choke up its channel. And then, where the torrent breaks impatiently from the lower end of another lochan, among hardened beds of Silurian grit and shale, we enter upon a great mass of granite, which forms the remaining mile of the course of the Girvan, and rises high on either hand into gray rugged hills. Crags of granite of every size and form stand up bleached and barren from the brown heath. Blocks of granite in endless varieties of bulk and shape lie strewed about, beneath and around the crags from which they have been detached. The river issues from a little tarn, called Loch Girvan Eye, filling a rock-basin in the granite, 1600 feet above the sea. Round this sheet of water the rugged ground is cumbered with blocks that seem just waiting their turn to be borne away down to the lower grounds. To the south, a high bleak mountain ridge ascends to an elevation of 2700 feet above the sea and 1100 over the parent tarn of the Girvan. Here, then, at last, is the source of the granite boulders of the valley. It was from these lonely hillsides that the Baron's Stone of Killochan was carried.

From these high grounds millions of boulders of all sizes, up to masses weighing at least thirty or forty tons, have been borne seawards and strewed over the lower hills and valleys of Carrick. What agency could transport them? It is plain that no flood of fresh water could have scattered them, for they are often perched on the hill-tops 800 or 900 feet above the valleys in which the streams are running. Nor is it conceivable that at a former time, when the level of the land was much lower than it is now, any great ocean-wave could have taken its rise within a limited area of what is now the highest ground in the south of Scotland, and carried with it in one vast resistless debacle such enormous quantities of boulders, so as not merely to bring them down into deep confined valleys, but actually to sweep them up again to the summits of the seaward hills.

Such work as this could have been done by only one agency in nature—that of ice.

When we once embrace the idea that the transport of these endless heaps of boulders has been effected by ice, the difficulties which previously seemed insuperable one by one disappear. And the more we examine into the facts of the case, the more firm becomes our conviction that this, after all, is the true theory. Looking at the Carrick hills with an eye that has been trained in the study of what are known as glacial phenomena, the geologist sees at every turn traces of a time when one wide mantle of ice and snow was thrown far and wide over the hills and valleys. The peculiarly-shaped hummocks and bosses of rock, so shorn and smoothed, recall at once the *roches moutonnées*, or ice-worn rocks, of Alpine valleys. The huge blocks of granite strewed along the hillsides remind one of the *blocs perchés* that abound on the flanks of the Swiss mountains, where they have been left by the retreating glaciers. The mounds of earth and rubbish, noted in the ascent of the course of the Girvan, are quite comparable with the moraines or rubbish-heaps that are shed from the ends of glaciers at the present day. Indeed, the whole contour of the ground, especially in the upper parts of the Girvan valley, suggests at a glance the former existence there of a massive sheet of ice which, descending ceaselessly from the higher tracts towards the sea, ground down and smoothed the surface of the rocks over which it moved. I have noticed in these uplands many examples of what are known as "dressed surfaces" on the rocks, and they are well seen in many places near the sea. These "dressings" are long ruts, scratches, and fine striæ, running in a determinate line across the smoothed surfaces of the rocks. They look like what might be artificially produced by pushing sand, gravel, and stones, under enormous pressure, along a polished plane of rock. And there cannot be any doubt that it was really by the attrition of such materials that the scratches were made, and that the pressure and onward movement were given by the vast overlying bed of ice. Similar dressings are familiar features of the rocks in Alpine valleys, where the trend of the striæ runs in the same line as the valley—that is, of course, in the direction in which the glacier has moved.

The water which percolates through the numerous joints and fissures of a rocky cliff and freezes there in winter, widens by its expansion the crevices it occupies. This operation being often repeated, there comes at last a time when the wedges of ice have so effectually sundered a mass from its parent cliff that it falls headlong into the valley. Should a glacier occupy the bottom of the valley below, the loosened rocks gather in heaps on the surface of the ice. Once there, they are slowly and steadily carried down the valley until—unless some rent in the ice should swallow them up by the way—they are thrown down at the end of the glacier, perhaps many leagues from the cliffs whence they originally came. In high northern latitudes the glaciers, instead of melting far in the interior of the country, as those of the Alps do, actually push their way out to sea, and break off in vast masses, which float away seaward as icebergs. It is clear that, if the surface of the glacier has been cumbered with boulders and rocky rubbish in the inland glens, it will carry this burden with it as it moves down to the sea-level; and the masses of ice which break off from the end of the glacier will, in like manner, bear their cargoes of earth and stones as they journey over the ocean. And, as these ice-islands melt away, their rocky cargoes must be scattered far and wide over the bottom of the sea. By this system of transport the ruins of many an Arctic valley are strewn over the fjords and sounds of Greenland.

At the time when the granite boulders of Carrick were transported from their original home among the hills, the land was so deeply buried under snow and ice that a massive ice-sheet crept down to the sea-level from the mountains of Carrick and Galloway, filling up the valleys and overriding the lower hills, even up to a height of more than 1000 feet above the present sea-level. The more precipitous eminences of the uplands rose above the surface of the ice on which they shed their frost-broken boulders of granite. Not improbably at the time of extremest cold the ice-sheet descended to the sea, and may have advanced for some way into its waters, where its margin broke up into fleets of bergs that sailed away seawards, dropping over the submerged land their freight of granite boulders. As happens within the Arctic circle at the present day, the cold may have been so intense as to freeze the waters of the ocean and invest the coast-line of that ancient Scotland with a solid encrusting zone of ice. Such an ice-cake would envelop many a stone lying along the beach, and, when broken up by the storms of summer, would carry its imprisoned boulders away to sea, and finally drop them on the bottom. It is far from improbable that this process was also in play during the long migration of the Carrick boulders. There still exist, in abundance, along some parts of the shores of the Clyde estuary, the remains of the shells which tenanted the sea during this cold era in our country's past history. Many of these shells are still natives of the neighbouring firth; some, however, and these often the most abundant, have long since died out in the British seas, though they still flourish in the waters of the Arctic Ocean. They are naturally adapted to a cold climate; and their abundance in the old sea-bottoms of the glacial period that occur on the west coast, affords a curious corroboration of the testimony of the boulders that the climate of the British Islands was once as severe as that of modern Greenland.

So here at last is the history of the origin of the Baron's Stone of Killochan. It once formed part of a cliff, some 2000 feet over its present site, far away up among the lonely mountains that look down upon Loch Doon. And, when it occupied its place in that cliff, the mountains around were cased deep in snow, and the glens were clogged with thick masses of ice which, with block-covered surface, moved steadily seaward. The granite cliff, like its representatives at the present day, traversed in all directions with joints and fissures, was liable to be split up into large angular blocks. One of these masses, weighing at least thirty-seven tons, was loosened one day from its resting-place and rolled down among the ruin of boulders that lay heaped upon the glacier below. With the ice in its steady seaward progress, this granite boulder moved mile after mile over ice-buried hill and glen; receiving, doubtless, many a dint from brother blocks hurried from their long silence in the cliffs to join the rattle of the ice-borne heaps beneath. Whether the transport was entirely done by the sheet of moving land-ice, or whether the last part of the journey was performed upon a detached berg floating off into the sea, may be matter of debate. But this at least is certain, that, after travelling some eighteen miles from its source, the boulder was finally stranded on or near the spot where it still remains. Many a shifting scene has come over the face of the country since then. The ice-fields have disappeared, and with them the hairy elephants and woolly rhinoceroses, reindeer and elks, which then roamed over the land; forests have sprung up and departed; the river has worn its way through cliffs of solid stone, and has rolled out many a fair meadow: but there still stands the granite boulder—a silent memorial of the long-vanished ice age.

But the Baron's Stone has another history, and from this it takes its name. The granite boulders of Carrick have served as an inexhaustible quarry from the earliest times. They may be seen forming a part of the ramparts of the hill forts of the early British tribes. Set upright, they sometimes have served as rough unchiselled monumental stones. A rude carving may, indeed, be traced on some of these monoliths. Thus, on the eastern flanks of the Brown Carrick Hill, a few miles south of the town of Ayr, lies an oblong block of gray granite weighing about two tons. It has evidently at one time been upright, and on the original face, which forms now the upper surface of the stone, a rude cross has been carved, having the same outline as the common monumental crosses of the West Highlands. That the stone served as a memorial of the dead can hardly be doubted. So simple an explanation, however, suited not the marvel-loving fancy of the old Carrick men. Abercrummie, the episcopal curate of Maybole, who was "outed" on the re-establishment of Presbyterianism, wrote a "Description of Carrick" about the close of the seventeenth century; and, in alluding to this sculptured stone, he calls it "a big whinstone, upon which there is the dull figure of a Crosse; which is alledged to have been done by some venerable churchman, who did mediat a peace twixt the King of the

Picts and Scots; and to give the more authority to his proposall did in their sight, by laying a crosse upon the stone, imprint that figure thereon." Another legend represents the cross as the impression of Sir William Wallace's sword, which, having been laid on the stone at nightfall, left its mould in the hard granite ere morning. A third version of the story relates how Wallace fought single-handed against a host of Englishmen, and how his sword, happening to strike against the stone, cut its likeness thereon by the blow!

The barons of Carrick found the boulders too hard to be dressed for the walls of their castles; but they used them with great effect to form the foundations, as in the stately castle of Dalquharran, on the banks of the Girvan. In recent times, as already said, they have been built into stone fences, cut into gateposts, and squared into blocks, of which tombstones and obelisks have been made.

The Baron's Stone of Killochan, however, does not seem ever to have had a tool upon it, until, some years ago, the proprietor had its name carved on its side to mark it as sacred from the hands of the relentless farmer. Tradition tells that it served as the judgment-seat of the old barons of Killochan, where they mustered their men, planned their raids, shared the booty, and hanged or cut off the heads of refractory prisoners. The family name is Cathcart, and the property still remains in their hands. They are said to trace their genealogy back to the days of the Bruce, a charter from whom still exists among the family archives. Though overshadowed by the power and influence of the Kennedies, the Cathcarts played their part in the troublous history of Carrick. Three brothers, including the Laird himself, died on the field of Flodden. Alan, third Lord Cathcart, fell at Pinkie. The son of the Flodden hero con-

trived to rouse the enmity of a branch of the Kennedies who had lands among the hills to the south, and suffered the loss of his left hand, besides sundry cuttings and woundings about the face. His grandson makes a more notable figure in the history. Choosing a pretty reach of the Girvan, a few hundred yards east from the Baron's stone, where possibly an older castle stood, he built a quaint mansion on the banks of the river, which still stands, and is known as the old House or Castle of Killochan. It is a characteristic specimen of the Scottish architecture of the period—a sort of passage from the old feudal keep or tower to the more recent mansion-house. The need of a strongly-fortified retreat, with loopholes and portcullis, had ceased to exist; but the builders still made their walls four or five feet thick, and, though they were no longer afraid to open out windows, they kept such openings as small as might be. They had been building flanking-towers so long too, that they could not but add one or two to the corners of the house. Moreover, they must needs cut the coping into embrasures, but instead of leaving them free for harquebuss or crossbow, they peaceably surmounted each with a short dumpy spire, like the cap of a pepper-box. Over the doorway is another indication of the advancing civilisation of the time; it is an inscription which runs thus:—
"This work was begun the 1 of Marche 1586 Be Johne Cathcart of Carlton and Helene Wallace his Spous The name of the Lord is ane strang tour and the rychteous in thair troublis rinnis into and findeth refuge. Prov 18 vers 10." It is unnecessary to remark that this is from an older translation of the Scriptures than our Authorised Version. The house—as appears from a curious set of carvings inside, representing the founder with his wife, and apparently his son and daughter—took several years to build. It stands

at the edge of a flat strip of alluvial meadow bordering the river, and is surrounded with old trees and hedgerows, and a terraced garden of the antique type. A year or two after the completion of his architectural and horticultural labours at Killochan, the Laird was summoned to attend "the Leutennentis Raid of Dumfreis." Like a great many other lairds, he thought proper to stay away, and was "delatit" in consequence. Next year—namely, at the close of 1601 —he was engaged, and his son with him, in one of the most memorable feuds in Ayrshire. The Laird of Bargany and the Earl of Cassilis, both Kennedies, and both comparatively young men, had long been at feud. Each jealous of the other's power, they were ready to fly to arms to avenge a real or fancied insult, and it cost King James no little anxiety to keep the peace between them. We find at one time the young Laird of Killochan sent by Bargany, his neighbour, to demand from the Earl of Cassilis the origin of a calumnious statement made by him. On another occasion, when there was like to be blood spilt between the rivals and their followers about the rents of certain fields near the sea, the old Laird Cathcart became surety for the peaceable settlement of the dispute. But these repeated quarrels, though quieted for a time, left their dark sediment of malice and revenge in the breasts of both the chieftains. "The King gart thame schaik handis," says the old chronicler of these feuds, "but not with their hairttis." At last, at the end of the year 1601, the Earl hearing that Bargany, with a small band of friends and retainers, was on his way south from Ayr, assembled a large armed force to waylay him. The two parties met near Maybole; Bargany, seeing the enormous disparity of numbers, tried to avoid a combat, and rode on with one part of his horsemen, while the young Cathcart followed at the head of the rest. But the Earl and his company were determined to use their advantage, and began to fire across the valley. Bargany's men being now in danger, he boldly rode forward with only two or three friends, and, pushing into the heart of his enemies, called out loudly for the Earl. Fighting his way onward, he soon had a host before and behind him. After a brave resistance, he was mortally wounded; but his horse bore him back to his own men, among whom he died soon after. The chronicler does not say what part the young Laird of Killochan took in the fight. He mentions the names of four comrades who dashed with Bargany into the ranks of the enemy, but Cathcart is not among them.

The next hundred years saw the reign of the Charleses and the Revolution, with the weary warfare of religious intolerance between Presbytery and Episcopacy. Ayrshire was a stronghold of the Presbyterians, and its remoter hills served as a favourite retreat from the authority of the Government. The old laird who built the house of Killochan must have witnessed the earlier scenes of that long strife, for he was alive towards the close of 1612, and in October of that year, "being sick in bodie, but haill in mynd," he made his will. He seems to have been in old age imbued with a large measure of the religious fervour of the period, if the words of Wodrow, as is probable, are to be referred to this individual. "The old laird of Carltoun was extraordinary at solving of cases of conscience," says Wodrow, and he gives an instance of how Dickson, who afterwards became a leader among the Presbyterians, had his doubts and fears as a student cleared away by the graphic exhortations of the old laird to whom he applied for relief. "The said Laird of Carltoun," he adds, "was wonderfully holy and heavenly in his family, and he had this peculiar way: He retired awhile his lone, be with him who would, before family worship, which ordinarily was before dinner, and came directly out of his closet to worship; and, be in the family who would, he retired immediately after worship to his closett till the meat was set on the table, and then he came to dinner and was extremely pleasant, for ordinary, to his conversation."

Some of the later lairds of Killochan have been in the army; but, though they have lived little on their estates in this part of Scotland, they have, with praiseworthy reverence, maintained the old house in its original condition. The wainscot fittings, thick-mullioned windows, old-fashioned grates, chairs, and cabinets, antique four-post beds and faded hangings, with the quaint grouping of tree and terrace, and mossy lawn round the building, still remain much as they were during the lifetime of the builder. Nor have they with less care guarded the oldest of all their heirlooms; and so, while the progress of agriculture has ploughed the fields, and swept away thousands of the huge granite boulders which of old cumbered the ground, the gentle green slope that looks down on the Girvan, and far away over to Ireland, still keeps its memories of the past, and its gray shattered Baron's Stone of Killochan.

IV.

THE COLLIERS OF CARRICK.[1]

COMPARATIVELY few of the many hundreds of tourists who flock every summer to that part of Scotland which the guide-books have styled "The Land of Burns" find their way farther south than "Alloway's auld haunted kirk" and the famous "brig" which lay so opportunely in Tam o' Shanter's line of retreat. When the weather is clear they get a distant view of the hills, which rise beyond the Doon into a background that has neither any striking outlines nor sufficient loftiness to form a notable feature in the remoter landscape. And yet if the visitor whose time and route are at his own disposal will bravely penetrate these far uplands, he will find much, both in the way of scenery and of historic and legendary interest, to reward his enterprise. It is a lonely pastoral region, deeply trenched with long and narrow valleys, the seaward portions of which are often well wooded and contrast with the singularly bare though verdant aspect of the high grounds on either side. The whole of that district was called in old times Carrick —a Celtic name still in use among the people, and descriptive of the rugged, rocky character of most of the surface. The bones of the country seem indeed everywhere to be sticking through the scanty skin of soil and turf; and yet

[1] *Good Words*, May 1873.

the abundant droves of black-faced sheep and black cattle, and the stores of excellent butter and cheese which every year come out of these hills to the great markets, bear witness to the quality of the pasture. It might have been hoped that in so rocky a tract minerals of some sort would be found to compensate for the comparative poorness of the surface. Many a viewer and "prospector" has scoured the sides of the hills and valleys. Copper, lead, and iron in small quantities have been found; but there seems no probability that the pastoral character of the country will ever be to any serious extent disturbed by mining operations. And yet, curiously enough, in one of the deep valleys on the northern margin of the hilly tracts of Carrick a small coal-field exists — a little bit of the great Scottish coal-field, which by some ancient terrestrial revolution has got detached from the rest, and become, as it were, jammed in between the two steep sides of the valley of the Girvan.

The colliers of Scotland have been in all time a distinct and a superstitious population. For many a long century they and the makers of salt were slaves, bought and sold with the land on which they were born, and from which they had no more right to remove themselves than if they had been of African descent, and born in Carolina. Customs and beliefs which had gradually died out elsewhere naturally lingered for a time among the colliers; and indeed until the general use of steam machinery and the invasion of an Irish labouring population, the Scottish miners maintained much of their singularity. Down in that little coal-field of Carrick, however, shut out from the rest of the mining districts, and even in no small degree from the country at large, the colliers preserved until only a few years ago many traits which we are accustomed to think had died out several generations before. No railway came near the place; no highway led through it. Lying near the sea, it yet could boast of no good harbour within reach to stimulate the coal industry. Even the local demand for coal was too small to admit of any extensive workings; and so the mining population continued the same quaint old ways which it had been used to for a century or two, keeping up, among other things, many of its characteristic superstitions.

Some years ago, on geological errand bent, I had occasion to pass a number of months in that sequestered locality, and to mingle with the colliers themselves, as well as their employers. In this way I was led to glean reminiscences of habits and beliefs, now nearly as extinct as the fossils in the rocks which were the more special objects of research. These gleanings, as illustrating former phases of our rural population, are perhaps not unworthy of record. I propose, therefore, in the present paper to relate an incident, perhaps one of the most striking in the history of coal-mining in this country, which occurred in this little Girvan coal-field, and which furnishes examples of several of the more characteristic features of the old Scottish collier.

In the quiet churchyard of Dailly, within hearing of the gurgle of the Girvan and the sough of the old pines of Dalquharran, lie the unmarked graves of generations of colliers; but among them is one with a tombstone bearing the following inscription :—

IN MEMORY OF
JOHN BROWN, COLLIER,
who was enclosed in
Kilgrammie Coal-pit, by a portion of it having fallen in,
Oct. 8th, 1835,
and was taken out alive,
and in full possession of his mental faculties,
but in a very exhausted state,
Oct. 31st,
having been twenty-three days in utter seclusion
from the world, and without a particle of food.
He lived for three days after,
having quietly expired on the evening of
Nov. 3rd,
Aged 66 years.

Three weeks without food in the depths of the earth! It seemed hardly credible, and I set myself to gather such recollections as might still remain. I discovered that a narrative of the circumstances had been published shortly after the date of their occurrence; but I was fortunate enough to make the acquaintance of people who were resident in the district during the calamity, and from whom I obtained details which do not seem ever to have found their way into print. Much of my information was derived from an old collier who was one of the survivors. His narrative and that of the other contemporaries of the event brought out in a strong light the superstition of the colliers, and furnished additional evidence as to one of the longest survivals without food of which authentic record exists.

On the 6th October 1835, in a remote part of the old coal-mine of Kilgrammie, near Dailly, John Brown, the hero of this tragedy, was at work alone. Sixty-six years of age, but hale in body and fond of fun, he had long been a favourite with his fellow-workmen, more especially with the younger colliers, whom his humour and story-telling used to bring to his side when their own term of work was done. Many a time would they take his pick from him, and finish his remaining task, while he sat on the floor of the mine, and gave them his racy chat in return. On the day in question he was apart from the others, at the far end of a roadway. While there, an empty waggon came rumbling along the rails, and stopped within a foot of the edge of the hole in which his work lay. Had it gone a few inches farther, it would have fallen upon him, and deprived him either of limb or life. There seemed something so thoughtless in such an act as the pushing of a waggon upon him that he came up to see which of his fellow-workmen could have been guilty of it. But nobody was there. He shouted along the dark mine; but no sound came back, save the echo of his own voice. That evening, when the men had gathered round the village fires, the incident of the waggon was matter of earnest talk. Everybody scorned the imputation of having, even in mere thoughtlessness, risked a life in the pit. Besides, nobody had been in that part of the workings except Brown himself. He fully acquitted them, having an explanation of his own to account for the movements of the waggon. He had known such things happen before, he said, and was persuaded that it could only be the devil, who seemed much more ready to push along empty hutches, and so endanger men's lives, than to give any miner help in pushing them when full.

In truth, this story of the waggon came in the end to have a significance little dreamt of at the time. It proved to have been the first indication of a "crush" in the pit— that is, a falling in of the roof. The coal-seam was a thick one, and in extracting it massive pillars, some sixteen or seventeen feet broad and forty to fifty feet long, were left to keep the roof up. At first, half of the coal only was taken out; but after some progress had been made the pillars were reduced in size, so as to let a third more of the seam be removed. This, of course, was a delicate operation, since the desire to get as much coal out of the mine as possible led to the risk of paring down the pillars so far as to make them too weak for the enormous weight they had to bear. Such a failure of support leads to a "crush." The weakened pillars are crushed to fragments, and at the same time the floor of the pit, under the enormous and unequal pressure, is here and there squeezed up even to the roof. Such was the disaster that now befell the coal-pit of Kilgrammie. It had been the early disturbance of level heralding the final catastrophe that sent the empty waggon along the roadway.

For a couple of days cracks and grinding noises went on continuously in the pit, the levels of the rails got more and more altered, and though the men remained at work, it became hourly more clear that part of the workings would now need to be abandoned. At last, on the 8th October, the final crash came suddenly and violently. The huge weight of rock under which the galleries ran settled down solidly on them with a noise and shock which, spreading for a mile or two up and down the quiet vale of the Girvan, were set down at the time as the passing of an earthquake. Over the site of the mine itself the ground was split open into huge rents for a space of several acres, the dam of a pond gave way, and the water streamed off, while the horses at the mouth of the pit took fright, and came scampering, masterless and in terror, into the little village, the inhabitants of which rushed out of doors, and were standing in wonderment as to what had happened.

But the disasters above ground were only a feeble indication of the terrors underneath. Constant exposure to risk hardens a man against an appreciation of his dangers, and even makes him, it may be, foolhardy. The Kilgrammie colliers had continued their work with reckless disregard of consequences, until at last the cry arose among them that the roof was settling down. First they made a rush to the bottom of the shaft, in hopes of being pulled up by the engine. But by this time the shaft had become involved in the ruin of the roof. A second shaft stood at a little distance; but this too they found to be closed. Every avenue of escape cut off, and amid the hideous groanings and grindings of the sinking ground, the colliers had retreated to a part of the workings where the pillars yet stood firm. Fortunately one of them remembered an old tunnel or "day-level," running from the mine for more than half a mile to the Brunston Holm, on the banks of the Girvan, and made originally to carry off the underground water. They were starting to find the entrance to this tunnel when they noticed, for the first time, that John Brown was not among them. Two of the younger men (one of whom told me the story) started back through the falling part of the workings, and found the old man at his post, working as unconcernedly as if he had been digging potatoes in his own garden. With some difficulty they persuaded him to return with them, and were in the act of hurrying him along, when he remembered that in his haste he had left his jacket behind. In vain they tried to drag him along. "The jacket was a new one," he said; "and as for the pit, he had been at a crush before now, and would win through it this time too." So, with a spring backwards, he tore himself away from them, and dived into the darkness of the mine in search of his valued garment. Hardly, however, had he parted from them when the roof between him and them came down with a crash. They managed to rejoin their comrades; John Brown was sealed up within the mine, most probably, as they thought, crushed to death between the ruins of the roof and floor.

Those who have ever by any chance peeped into the sombre mouth of the day-level of a coal-pit will realise what the colliers had now to do to make good their escape. The tunnel had been cut simply as a drain; dark water and mud filled it almost to the roof. For more than half a mile they had to walk, or rather to crouch along in a stooping posture through this conduit, the water often up to their shoulders, sometimes, indeed, with barely room for their heads to pass between the surface of the slimy water and the rough roof above. But at length they reached the bright daylight as it streamed over the green holms and autumn woods of the Girvan, no man missing save him whom they had done their best to rescue. They were the first to bring the tidings of their escape to the terrified village.

No attempt could at first be made to save the poor prisoner. As the colliers themselves said, not even a creel or little coal-basket could get down the crushed shaft of the pit. The catastrophe happened on a Wednesday, and when Sunday came the parish minister Dr. Hill—afterwards a conspicuous man in the Church of Scotland—made it the subject of a powerful appeal to his people. In the words of a lady, who was then, and is still, resident in the neighbourhood, "he made us feel deeply the horror of knowing that a human being was living beneath our feet, dying a most fearful death. On the Sunday following we met with the conviction that whatever the man's sufferings had been, they were at last over, and that he had been dead some days. On the third Sunday the event had begun to pass away."

After the lapse of some days the cracking and groaning of the broken roof had so far abated that it became possible once more to get down into the pit. The first efforts were, of course, directed towards that part of the workings where the body was believed to be lying. But the former roadways were found to be so completely blocked up that no approach to the place could be had save by cutting a new tunnel through the ruins. This proved to be a work of great labour and difficulty; for not only were the materials extremely hard through which the new passage must be cut, but an obstacle of another kind interrupted the operations —a dead body lay in the pit, and awakened all the superstition of the colliers. At times they would work well, but their ears were ever on the alert for strange weird noises, and often would they come rushing out from the working in terror at the unearthly gibberings which ever and anon would go soughing through the mine.

A fortnight had passed away. The lessee, like the rest of the inhabitants, believed poor Brown to be already dead, and brought a gang of colliers from another part of the county to help in clearing out and reopening his coal-pit. But a party of the men continued at work upon the tunnel that was to lead to the body. They cut through the hard crushed roof a long passage, just wide enough to let a man crawl along it upon his elbows; and at last, early on the morning of the twenty-third day after the accident, they struck through the last part of the ruined mass into the open workings beyond. The rush of foul air from these workings put out their lights, and compelled them to retreat. One of their number was despatched to upper air for a couple of boards, or corn-sieves, or any broad flat thing he could lay hands upon, with which they might advance into the workings, and waft the air about, so as to mix it, and make it more breathable. Some time had to elapse before the messenger could make the circuitous journey, and meanwhile the foulness of the air had probably lessened. When the sieves came one of the miners agreed to advance into the darkness, and try to create a current of air; the rest were to follow. In a minute or two, however he rejoined them, almost speechless with fright. In winnowing the air with his arms, he had struck against a waggon standing on the roadway, and the noise he had made was followed by a distinct groan. A younger member of the gang volunteered to return with him. Advancing as before, the same waggon stopped them as their sieves came against the end of it, and again there rose from out of the darkness of the mine a faint but audible groan. Could it be the poor castaway, or was it only another wile of the arch enemy to lure two colliers more to their fate? Gathering up all the courage that was left in him, one of them broke the awful silence of the place by solemnly demanding, "If that's your ain groan, John Brown, in the name o' God, gi'e anither." They listened, and after the echoes of his voice had ceased they heard another groan, coming apparently from the roadway only a few yards ahead. They crept forward, and found their companion—alive.

In a few seconds the other colliers, who had been anxiously awaiting the result, were also beside the body of John Brown. They could not see it, for they had not yet ventured to rekindle their lights; but they could feel that it had the death-like chill of a corpse. Stripping off their jackets and shirts, they lay with their naked backs next to him, trying to restore a little warmth to his hardly living frame. His first words, uttered in a scarcely audible whisper, were, "Gi'e me a drink." Fearful of endangering the life which they had been the means of so marvellously saving, they only complied so far with his wish as to dip the sleeve of a coat in one of the little runnels which were trickling down the walls of the mine, and to moisten his lips with it. He pushed it from him, asking them, "no to mak' a fule o' him." A little water refreshed him, and then, in the same strangely sepulchral whisper, he said, "Eh, boys, but ye've been lang o' coming."

Word was now sent to the outer world that John Brown had been found, and was yet living. The lessee came down, the doctor was sent for, and preparations were made to have the sufferer taken up to daylight again. And here it may be mentioned that upon the decayed timber props and old wooden boardings of a coal-pit an unseemly growth of a white and yellow fungus often takes root, hanging in tufts and bunches from the sides or roofs wherever the wood is decaying. After being cautiously pushed through the newly-cut passage, John Brown was placed on the lessee's knees on the cage in which they were to be pulled up by the engine. As they rose into daylight, a sight which had only been faintly visible in the feeble lamplight below presented itself, never seen before and never to be forgotten. That coal-mine fungus had spread over the poor collier's body as it would have done over a rotting log. His beard had grown bristly during his confinement, and all through the hairs this white fungus had taken root. His master, as the approaching daylight made the growth more visible, began to pull off the fungus threads, but (as he told me himself) his hand was pushed aside by John, who asked him, "Na, noo, wad ye kittle [tickle] me?"

By nine o'clock on that Friday morning, three-and-twenty days after he had walked out of his cottage for the last time, John Brown was once more resting on his own bed. A more ghastly figure could hardly be pictured. His face had not the pallor of a fainting fit or of death, but wore a strange sallow hue like that of a mummy. His flesh seemed entirely gone, nothing left but the bones, under a thin covering of leather-like skin. This was specially marked about his face, where, in spite of the growth of hair, every bone looked as if it were coming through the skin, and his eyes, brightened into unnatural lustre, were sunk far into his skull. The late Dr. Sloan, of Ayr, who visited him, told me that to such a degree was the body wasted that, in putting the hand over the pit of the stomach, one could distinctly feel the inner surface of the backbone. Every atom of fatty matter in the body seems to have been consumed.

Light food was sparingly administered, and he appeared to revive, and would insist on being allowed to speak and tell of his experiences in the pit. He had no food with him all the time of his confinement. Once before, when locked up underground by a similar accident, he had drunk the oil from his lamp and had thereby sickened himself; so that this time, though he had both oil and tobacco with him, he had tasted neither. For some days he was able to walk about in the open uncrushed part of the mine, where too he succeeded in supplying himself with water to drink. But in the end, as he grew weaker, he had stumbled across the roadway and fallen into the position in which he was found. The trickle of water ran down the mine close to him, and was for a time the only sound he could hear, but he could not reach it. When asked if he had not despaired of ever being restored to the upper air, he assured his questioners that he had never for a moment lost the belief that he would be rescued. He had heard them working towards him, and from the intervals of silence and sound he was able, after a fashion, to measure the passing of time. It would seem, too, that he had been subject either to vivid dreams or to a wandering of the mind when awake, for again and again he thanked the sister of his master for her great kindness in visiting him in the pit and cheering him up as she did.

On the Sunday afternoon, when some of his old comrades were sitting round the bedside, he turned to them with an anxious puzzled look and said, "Ah, boys, when I win through this, I've a queer story to tell ye." But that was not to be. His constitution had received such a shake as even its uncommon strength could not overcome. That evening it became only too plain that the apparent recovery of appetite and spirits had been but the last flicker of the lamp of life. Later in the night he died.

So strange a tragedy made a deep impression on the people of that sequestered district. Everybody who could made his way into the little cottage to see a man who, as it were, had risen from the dead; and no doubt this natural craving led to an amount of noise and excitement in the room by no means very favourable to the recovery of the sufferer. But this was not all. A new impetus was given to the fading superstitions of the colliery population. Not a few of his old workfellows, though they saw him in bodily presence lying in his own bed and chatting as he used to do, nay, even though they followed him to the grave, refused to believe that what they saw was John Brown's body at all, or at least that it was his soul which animated it. They had seen so many wiles of the devil below ground, and had so often narrowly escaped with their lives from his treachery, that they shrewdly suspected this to be some new snare of his for the purpose of entrapping and carrying off some of their number.

A post-mortem examination followed. But even that sad evidence of mortality failed to convince some of the more stubbornly superstitious. The late Dr. Sloan, who took part in the examination, told me that after it was over, and when he emerged from the little cottage, a group of old colliers who had been patiently waiting the result outside came up to him with the inquiry, "Doctor, did ye fin' his feet?" It certainly had not occurred to him to make any special investigation of the extremities, and he confessed that he had not, though surprised at the oddity of the question. He inquired in turn why they should have wished the feet particularly looked to. A grave shake of the head was the only reply he could get at the time; but he soon found out that had he examined the feet, he would have found them not to be human extremities at all, but bearing that cloven character which Scottish tradition has steadily held to be one of the characteristic and inefface able features of the "deil," no matter under what disguise he may be pleased to appear.

And even when the grave had closed over the wasted remains of the poor sufferer, people were still seeing visions and getting warnings. His ghost haunted the place for a time, until at last the erection of a tombstone by the parishioners with the inscription already quoted, written by the parish minister, slowly brought conviction to the minds of the incredulous. Many a story, however, still lingers of the kind of sights and sounds seen as portents after this sad tragedy. I shall give only one, told to me by an old collier, whose grandmother was a well-known witch, and who himself retained evidently more belief in her powers than he cared to acknowledge in words. Not long after

John Brown's death, one of the miners returned unexpectedly from his work in the forenoon, and to the surprise of his wife appeared in front of their cottage. She was in the habit, unknown to him, of solacing herself in the early part of the day with a bottle of porter. On the occasion in question the bottle stood toasting pleasantly before the fire when the form of the "gudeman" came in sight. In a moment she had driven in the cork and thrust the bottle underneath the blankets of the box-bed, when he entered, and, seating himself by the fire, began to light his pipe. In a little while the warmed porter managed to expel the cork, and to escape in a series of very ominous guggles from underneath the clothes. The poor fellow was outside in an instant, crying, "Anither warning, Meg! Rin, rin, the house is fa'ing." But Meg "kenn'd what was what fu' brawly," and made for the bed in time to save only the last dregs of her intended potation.

Most of the actors in the sad story have passed away, and now rest beneath the same green sod which covers the remains of John Brown. With the last generation, too, has died out much of the hereditary superstition. For a railway now runs through the coal-field. Strangers come and settle in the district. An increasing Irish element appears in the population, and thus the old manners and customs are rapidly becoming mere traditions in the place. Even grandsons and great-grandsons of the old women who "kept the country-side in fear," affect to hold lightly the powers and doings of their progenitors, though there are still a few who, while seemingly half-ashamed to claim supernatural power for their "grannies," gravely assert that the latter had means of finding things out, and, though bedridden, of getting their wishes fulfilled, which, to say the least, were very inexplicable.

V.

AMONG THE VOLCANOES OF CENTRAL FRANCE.[1]

IT had been my good fortune to spend several years in a more or less continuous examination of those volcanic hills and crags which form so characteristic a feature in the scenery of the great central valley of Scotland. I had traced them over many hundreds of square miles, sometimes underneath the very streets and squares of a town, sometimes across richly-cultivated fields, and sometimes far inland among lonely moors and mosses. I had studied their association with the stratified rocks of that old era of this country's history known as the Carboniferous Period; I had thus been enabled, in some measure, to realise the scenery of that ancient time—its wide jungles and lagoons, crowded with graceful trees, and dotted here and there with dark pine-clothed volcanic cones that sent out their columns of steam and showers of ashes, or rolled their streams of lava into the shallow waters. My restorations of the Carboniferous landscapes, however, could not but be incomplete and unsatisfactory. They wanted spirit and life, even more than the plaster model of some extinct monster constructed from the hints that may be suggested by a tooth and a few bones. They needed comparison with some region of

[1] *Vacation Tourists and Notes of Travel in* 1861, Macmillan and Co.

recent volcanoes, where, like the dry bones in the field of old, they might straightway be touched into life.

As the Scottish volcanoes had been of small extent, as well as eminently sporadic in their distribution, it seemed to promise more success to compare them with a district where similar local phenomena had been manifested, than with such regions as those of Etna or Vesuvius, where the eruptions had been on a larger scale, and had proceeded from the different vents of one great volcano. There were two districts in Europe that appeared likely to throw light on the subject—one of these lay in the Eifel, the other in the high grounds of Auvergne and the Haute-Loire. The latter covers a much greater area than the German tract, and presents besides a more extensive variety of volcanic phenomena. It had been described in detail in the admirable volume of Mr. Poulett Scrope, as well as in several other works and memoirs by able geologists in France and England. These writings did not, indeed, treat the geological structure of the country from the particular point of view which chiefly interested me at the time, but they formed an invaluable guide to one who wished to acquire as rapidly as possible a general knowledge of the region. So it was resolved by an old comrade and myself to go to Auvergne, and enlarge our ideas in one department of British geology. Between two countries once so closely linked together in peace and war, it seemed as if there might be another relationship than that of mere State policy; and so with some such fanciful notion we set out to see how far we could succeed in establishing a geological connection between Central Scotland and Central France.

Not many years ago it was a matter of no little discomfort to reach the high grounds of the Puy de Dôme and the other departments in the interior of France. Several days of diligence travelling, and inns none of the best, were hindrances seldom surmounted save by enthusiastic geologists, or by valetudinarians who risked all peril to spend a few weeks at the Baths of Mont Dore. Now, however, this state of things has changed. Railways penetrate far into the upland districts, and although this part of France is still comparatively little known to English tourists, it can be visited with even more ease, and in a shorter time, than the remoter parts of Scotland. Dining on a summer evening in London, one may take one's seat in the Dover express about nine o'clock, and next evening at the same hour may see the sun set behind the long chain of *puys* that dot the granitic plateau of Auvergne. The journey from Paris southward, indeed, is a dreary and monotonous one, even if you make it at the rate of thirty-five or forty miles an hour. Wide uninteresting plains occupy hundreds of square miles, and it is not until towards the close of the day, as you approach the department of Allier, that the ground begins to undulate, amid hedgerows of acacias and patches of woodland. From the quaint old town of Moulins the scenery becomes hourly more interesting. A vast, richly-cultivated plain, several miles broad, and known as the Limagne d'Auvergne, widens out southward and stretches as far in that direction as the eye can reach. On the east lies the chain of granite hills which separates the plain of the Allier from the basin of the Loire, while to the west the eye rests with increasing wonder upon a long line of conical hills, sometimes bare and gray, sometimes dark with foliage, and grouped like a series of colossal forts and earthworks along the summit of a long ridge. Beyond these, and seemingly rising out of them, towers the grand cone of the Puy de Dôme, now flushed, perhaps, with the last rays of the sinking sun. As the train advances southward these cones become still more defined, standing up dark and clear against the evening sky, until, halting at last at Clermont we seem to rest almost at the feet of the giant Puy.

The ancient province of Auvergne—now parcelled out into the departments of Cantal, Puy de Dôme, and Haute-Loire—comprises a considerable part of the high ground in Central France, and from the variety of its geological structure contains a diversity of outline that contrasts well with the monotonous scenery of so much of the lower parts of the country. Granite and other crystalline rocks rise from under encircling plains of Secondary and Tertiary strata, and form an elevated tableland in the central districts, through which run the valleys of the Loire, the Allier, the Dore, the Sioule, and other minor rivers.

At a comparatively recent geological period there were some large lakes in these uplands, one of them extending over the modern Limagne d'Auvergne in a north and south direction, between granitic hills, for a distance of fully forty miles, and with a breadth of sometimes twenty. But the lakes have long since disappeared, though their site is still marked by broad plains formed of lacustrine strata, often composed of the remains of the shells that lived in these inland waters. It was in this region of high ground, among hills of granite, gneiss, and schist, watered by large rivers and by broad lakes, that those volcanic eruptions broke forth, to some of whose features it is the object of the present paper to direct attention. To such protrusions of igneous matter the great altitude of some parts of the district is due. Lava and ashes have been thrown out upon the granitic hills, so as to rise even into great mountains, where, as in the higher and deeper recesses of Mont Dore, snow may be seen gleaming white among the crags under the glare of a July sun.

The easiest point from which to begin the examination of this region is probably Clermont, the chief town of the department of Puy de Dôme. Built round a small hill on the west side of the Limagne, where that broad valley attains its greatest width, Clermont rises conspicuously above the general level of the plain (which is about 1200 feet above the sea-level), and seems to nestle at the base of the long granitic ridge that supports the chain of Puys. The hill on which the town is placed is of volcanic origin, so too are similar gentle eminences that rise above the level country towards the east; north and south, at the distance of a mile or two, are remnants of ancient lava-beds, now forming flat-topped hills; while to the west, down some of the narrow gullies that descend through the granitic ridge, currents of lava have forced their way from the volcanic vents of the Puys almost to the very site of the town. Here, then, the traveller may rest for a while, with plenty of geological interest around him if he care to ply his hammer, and with not less of varied and curious scenery if he be only in search of the novel and picturesque. Let no man, however, whether geologist or not, visit Auvergne in July, unless fully prepared to eat, drink, and be merry, with the thermometer at 82° or more in the shade.

Our first geological ramble was begun soon after sunrise. Passing through a labyrinth of lanes and byways, we succeeded in reaching the base of the hills, and began to wind upwards among the vineyards that cluster along the slopes and look down upon the rich plain of the Limagne. It was a glorious morning. A light mist hung over the valley, concealing its features as completely as if the lake which once filled it had been again restored; while some twenty miles to the eastward, on the farther side of this sheet of phantom water, rose the purple hills of the Forez that separate the basins of the Allier and the Loire. Behind us, as we looked across the plain, lay the great granitic ridge or plateau, rising to a height of somewhere about 1600 feet above the plain, and nearly 3000 feet above the sea. Its base, up which we were slowly ascending, had a varied mantle of cornfields and vineyards; narrow, well-wooded valleys had been cut by streamlets down its flanks, but the higher slopes became barer by degrees as they approached the range of volcanic cones that crowns the summit of the ridge. It was with no slight interest that, among the little runnels and cart-tracks which were crossed in the ascent, we watched for indications of the nature of the rocks below. Sometimes a chalky lacustrine marl was noticed; and, as we drew nearer to the granite, we found ourselves upon pebbly sandstone that had evidently been formed out of the waste of the granite hills. But how could the formation of such a deposit have been effected here? Foot by foot as we crept up the acclivity this sandstone accompanied us, until at last, at a height of probably not less than a thousand feet above the level of the plain, we reached the granite. The gravel and sand, out of which this sandstone had been made, must have been deposited in a lake—the old lake, in short, which once occupied the site of the Limagne. The water must therefore have reached up as far as to the point to which we had traced the sandstone; and thus, in the course of an hour's ramble, we ascertained for ourselves the somewhat startling fact, that unless later subterranean movements had altered the relative levels, the fertile plain below was formerly covered by a lake at least a thousand feet deep. Once on the granite we were free from the entanglements of enclosures and fences. As this rock crumbles away with rapidity, its surface is smooth, without those rugged features which mark the surrounding basaltic rocks. It is coated with a short scrubby grass, save in those places where the amount of waste is too great and rapid to allow the vegetation to take root. Crossing a short interval of this ground, at the height of about 2400 feet above the sea, we arrived at the basalt that caps the ridge of Pradelle.

From this height we commanded a wide view of the Limagne, from which the morning sun had now dispelled the floating mists; we could judge better of the disposition of the volcanic cones, or puys, and of the aspect of some of the basaltic plateaux and lava-streams. But the most impressive part of the scene was not in the traces of old igneous eruptions, but in the evidence of the power of running water. I had wandered long among the basalt hills of the Hebrides, and now recognised the repetition of many features of their landscapes; but nothing I had seen or read of had prepared me for such a stupendous manifestation of the power of rain and rivers. No one, indeed, whose observations have been confined to a country which has been above the sea only since the glacial period, or the contours of which have been smoothed over by the ice-sheets of that time, can readily form an adequate idea of the denuding effect of water flowing over the surface of the land. Standing on the plateau of Pradelle, with its remnant of a lava-current, and looking down into the valley of Villar—a deep gorge, excavated by a rivulet through that lava-current, and partially choked up by a later *coulée* of lava which the stream is now wearing away—I received a kind of new revelation, so utterly above and beyond all my previous conceptions was the impression which the sight of this landscape now conveyed. The ridge of Pradelle is a narrow promontory of granite, extending eastward from the main granitic chain, and cut down on either side, but more especially to the south, by a deep ravine. It is capped with a cake of columnar basalt, which of course was once in a melted state, and, like all lava-streams, rolled along the ground ever seeking its lowest levels. A first glance is enough to convince us that this basaltic cake is a mere fragment, that its eastern and southern edges have been largely cut away, and that it once extended southwards across what is now the deep gorge of Villar. Since the eruption of the basalt, therefore, the whole of this gorge has been excavated. But what agent could have worked so mighty a change? We bethink us, perhaps, of the sea; and picture the breakers working their way steadily inland through the softer granite. But this supposition is untenable, for it can be shown on good grounds that, since the volcanic eruptions of this district began, the country has never been below the sea. It is with a feeling almost of reluctance that we are compelled to admit, in default of any other possible explanation, that the erosion of the valley has been the work of the stream that seems to run in a mere rut at the foot of the slopes. How tardy must be the working of such an agent, and how immeasurably far into the past does the contemplation of such an operation carry us! This illustration of the power of running water, however, though the first, was by no means the most striking which occurred in the course of my rambles in Auvergne. The same fact stood out with a kind of oppressive reality in the Haute-Loire, to which reference will be made on a subsequent page.

The basalt of Pradelle recalled many of the basaltic hills in various parts of Scotland. I could have supposed myself under one of the cliffs that look out upon the deep fjords of Skye, or below the range of crags on the shores of the Forth, over which Alexander III. lost his life, or even among some of the ridges that form the eastern part of Arthur's Seat, at Edinburgh. The French basalt had, indeed, a grayer colour and a finely cavernous structure, which distinguished it from the hard black compact rock which is known as basalt in Scotland: but they were columned both in the same way, traversed by similar transverse joints, and, above all, resembled each other in their mode of yielding to the weather, and in their general aspect in the landscape.

Quitting this ridge, and walking westwards towards the Puy de Dôme, we reached the hostelry of Bonabry, where the road splits into two, one branch crossing the hilly ground for Pont Gibaud, the other turning south-west for Mont Dore. Here, finding the morning too far advanced for further breakfastless exploration, we struck down for the valley of Villar, with the view of examining more narrowly a later current of lava in the bottom of the ravine—a barren expanse of black rugged scoriæ rising into the most fantastic forms, and nearly destitute of vegetation. This lava-current must be greatly more recent than that of Pradelle, for it has been erupted after the excavation of the valley. Few walks in Auvergne are in their way more instructive than this. The valley itself, with its impressive lesson of river-action, becomes still more striking when seen from below. The Pradelle basalt hanging over the ravine stands as a silent witness at once of the antiquity of the earlier volcanic eruptions and of the changes of after time. The great river of younger lava below, too, is an object of unceasing interest to a geological eye, winding as it does with all the curvings of the valley, now sinking down beneath a mass of tangled copsewood, and now rising up into black craggy masses, where some projecting boss of granite had formed a temporary impediment to its course. The rivulet has actually cut in places a second narrow gorge through the lava, sometimes of considerable depth. But part of the stream still appears to flow down the old channel beneath the lava by which that channel has been usurped, for at the abrupt termination of the lava-current an abundant gush of water issues from under the black rugged crags.

In the town of Clermont itself there is not much of interest. It is built round the sides of a gently-sloping hill, and thus the towers of the old church, rising to a considerable height above the surrounding plain, can be seen from a great distance. This church, like most of the rest of the town, is built of a dark, compact lava, that gives a somewhat sombre hue to the building. The same tone of colouring would also characterise the street architecture, but for a plentiful use of whitewash. One cannot but admire the sharpness with which this lava has retained for centuries its chisel-marks and sculpturings; even staircases, that have been trodden so long day after day, seem well-nigh as fresh as ever. So black and dingy, indeed, and so sharp in outline, are some of the tall pillars, that they might readily be mistaken for so many shafts of cast-iron. Along the roadsides, too, you constantly pass crosses made of the same material—black, sombre things, rising sometimes from the edge of a vineyard, sometimes standing up alone in a solitary part of the way, among broken walls and thickets of brushwood. It was not uninteresting to remember that some three hundred years ago the roadsides at home were studded with similar crosses, of which the pedestals and parts of the stems may still, here and there, be seen; and that these were in many cases made of an old lava, just as in Auvergne. The Scottish rock, however, had been erupted many a long geological period ere the Auvergne volcanoes broke forth; and though the crosses hewn out of it may not have dated further back than some of these French ones, yet Nature has dealt kindlier with them, crusting them over with lichen and moss, and making them look as crumbling and venerable as the crags and hillsides that rise around them. The Auvergne lava, on the other hand, is a singularly barren stone; it gives no harbourage to vegetation, and its chiselled surfaces stand up now as bare and blank as they have done for centuries.

No one should leave Clermont without looking at the baths of Saint Alyre. A spring, highly charged with carbonate of lime, issues from the side of the hill of Clermont, and deposits along its course a constantly-increasing mass of white travertin. In this way it has formed for itself a natural aqueduct, running for a considerable distance, and terminating in a rude but picturesque arch of the same material, below which flows a small stream. The water that trickles over this bridge evaporates, and leaves behind a thin pellicle of carbonate of lime, which gathers into rugged masses, or hangs down in long stone icicles or stalactites. Such a *fontaine pétrifiante* could not remain a mere curiosity: it has been turned into a source of considerable profit, and manufactures for the visitors an endless stock of brooches, casts, alto-relievoes, basso-relievoes, baskets, birds' nests, groups of flowers, leaves, fruit, and suchlike. A portion of the water is diverted into a series of sheds, where it is made to run over flights of narrow steps, on which are placed the objects to be "petrified." By varying the position of these objects, and removing them farther and farther from the first dash of the water, they become uniformly coated over with a fine hard crust of white carbonate of lime, which retains all the inequalities of the surface on which it is deposited. There is here, of course, no real *petrification*; the substances operated upon retain all their original structure, and are only *incrusted* with the calcareous sediment. When once covered with this stony crust, they may remain unchanged for a long period, being thus hermetically sealed and protected from the influences of the air.

Let the reader suppose himself on the top of the Puy de Dôme, 4842 feet above the sea-level. Seated on the greensward which covers that elevated cone, he has the volcanic district spread out as in a map below him—cones, craters, and lava-currents—clear and distinct for many miles to the north and south. The Puy de Dôme, placed about midway between the northern and southern ends of the chain of the Puys, rises out of the centre of the long granitic ridge or plateau on the western edge of the valley of the Allier. Its position, therefore, is eminently favourable for obtaining a bird's-eye view of the country. Below us, to the eastward, lies the broad plain of the Limagne like a vast garden, dotted here and there with hamlets and villages and towns. Yonder, for instance, are the sloping streets of Clermont, with their dingy red-tiled houses, and the sombre spires of the old church; farther eastward is Montferrand, and others of lesser note lie in the district beyond. The eastern horizon is bounded by the range of the granitic hills of the Forez, which have been already referred to as rising from the level of the Limagne on the one side, and descending into the basin of the Loire on the other. They look gray and parched in the glare of the summer afternoon, though softened a little by the purple light of distance, till their base seems to melt into the subdued verdure of the valley. Westward, the eye wanders over a dreary region of broken and barren ground which stretches far to the north, while southward, some fifteen or twenty miles away, it sweeps round into the mountains of Mont Dore that terminate the southern landscape.

It is the nearer prospect, however, which forms the chief source of wonder as we look from the summit of the Puy de Dôme. Between us and the great plain of the Limagne lies a strip of the elevated granitic plateau—a tract of bare uneven ground, traversed by some deep valleys that descend towards the east. On this plateau rises a chain of isolated conical hills, stretching due north and south from the Puy de Dôme, which is the highest point in the district. Unconnected by ridges and watersheds into a regular chain, like a common range of hills, they shoot up from a dark sombre kind of tableland, at a steep angle, into cones which seem to be completely separated from each other. Cone behind cone, from a mere hillock up to a good hill, rises from the brown waste for some twenty miles to the north and south of the great Puy. Some of them are partially clothed with beechwoods, but most have a coating of coarse grass and heath, intermingled here and there with numerous wild flowers. Where devoid of vegetation, their slopes consist of loose dust and stones, like parts of the tableland on which they stand. Wolves still harbour in their solitudes, among the dense woods that clothe some of the slopes, and the shepherds have to keep a good look-out after their flocks. At the top of the Puy de Dôme I found a boy, of ten or twelve years, armed with a club-headed staff, which he told me was used against the audacious wolves, and he pointed to a thick forest on a neighbouring hill whence the animals made their forays.

Not the least singular feature of these conical hills is, that nearly all of them look as if they had had their tops shaved off. Nay, they even seem in the distance to have been more or less scooped out, as if some old Titan had taken a huge spadeful out of the summit of each hill. The reason of this structure may be guessed, but it becomes strikingly apparent on a closer inspection of the ground. Each cone, with four or five exceptions, is found on examination to be an actual volcano, extinct indeed, but still well-nigh as fresh as if the internal fires had burnt out only yesterday. The truncated, hollowed summit thus turns out to be a true crater—the vent, in short, whence the materials of the hill were erupted. Upwards of fifty such volcanoes dot the ridge to the north and south of the Puy de Dôme, each formed from an independent orifice, and sometimes containing, as in the Puy de Montchié, no fewer than four separate craters in one hill. They consist of loose ashes, dust, and scoriæ, still so lightly aggregated that, where the rain has bared off long strips of the grassy covering, one may slide rapidly ankle-deep in *débris* from the top of a cone to its base. Many of the cones have had one of their sides removed, and from the broken part a current of basaltic lava has issued, flowing out over the tableland, sometimes for several miles, and even descending the valleys that slope into the Limagne. The main mass of lava, in many different streams, has gone down the western side of the chain towards the valley of the Sioule, and hence the strange, sombre, arid aspect of that tract. From the summit of the Puy de Dôme you can trace some of the lava-streams, marking whence they issued and how they flowed across the country. That of the Villar valley, already described, is especially noticeable, breaking from the Puy de Pariou, and descending towards the east in a black rugged current, like a river of frozen icebergs.

Such, then, is the general landscape that stretches around the great Puy de Dôme. It is eminently dreary and desolate in the nearer parts, while in the eastern distance the eye rests on the bright, corn-clad Limagne. The long line of volcanic cones stretching to the north and south affords every facility to the geologist, and presents him, moreover, with a class of phenomena not found round the larger active volcanoes of Europe. The independence, small extent, number, and local distribution of the cones are features that throw light on what must have been the character and aspect of the Carboniferous volcanoes of Central Scotland, to illustrate which had been the object of my visit to Auvergne. A closer examination of these cones brings out a further parallelism with the more ancient vents. The Puy de Pariou, for example—one of the most accessible, and at the same time one of the most perfect, cones of the chain—lies somewhat more than a mile due north of the Puy de Dôme. It consists, in reality, of two craters, but only a portion of the northern rim of the older one is now visible, the rest being occupied by the newer crater, which is still in a perfect state of preservation. Ascending, as is usual, from the east side, the visitor first passes over a lava-current. From the foot of the cone the ascent is tolerably steep, among coarse grass, violets, martagon lilies, yellow gentians, and many other flowers, until the top of the older cone is reached, whence he looks down into the first crater, with the gap which the lava-current has made in it. Walking southward along its rim, he sees it passing under a later cone, which reaches a height of 738 feet above the plateau from which the southern side of the hill rises. After a second ascent, he arrives at last at the top of the Puy, and finds that the newer cone has been erupted over the southern half of the older one, and that it contains a beautifully perfect crater. Hence, from the top of the Puy there is on the south side an unbroken declivity, sloping at about 35°, down to the surface of the tableland, while on the north side the inner cone descends first into the older crater, which half encircles it. The last-formed crater measures 3000 feet in circumference. It is an inverted cone; its sides are smooth and grassy, and shelve steeply down to a depth of 300 feet. They have been indented by a series of cattle-tracks, rising in successive steps above each other, which Mr. Scrope aptly compares to the seats of an amphitheatre. Nothing can be more complete or regular than this part of the Puy. While ascending the outer slopes, one looks forward to reach a broad flat tableland on the top, carpeted perchance with the same coarse heather and wild flowers as clothe the sides of the hill: but, instead of level ground, one gazes down into a deep, round, smooth-sided crater, covered with grass to the bottom. Between the inward slope of this hollow and the outward declivities of the Puy, the rim is at times so narrow that you may almost sit astride on it, one foot dangling into the crater, the other pointing down to the plateau from which the hill rises. And there, with wild flowers clustering around, butterflies hovering past, cattle browsing leisurely down the sides of the crater below, while the tinkle of the sheep-bells ever and anon comes up with the scented breeze from the outer slopes of the Puy, one cannot without an effort picture the turmoil and violence to which the Puy owes its rise, when the ground was rent by subterranean explosions, and when showers of dust and stones were thrown out from the orifice.

From the older crater, now more than half filled up by the last eruptions, a stream of lava passes out northwards, through a great gap in the cone, trending at once to the east, over the plateau and down the valley of Villar. Here the history of the whole Puy is at once apparent. First of all, after some underground movements, a fracture was made, through which gas, steam, ashes, and scoriæ were vomited forth. The ejected material fell back again, partly into the vent, partly round its margin, gathering by degrees into a cone with a crater in its centre. A column of lava rose in the vent, began to fill the bowl-like cavity of the crater, and continued to well upward until the loosely-compacted sides of the hill were no longer able to withstand the pressure of the increasing mass of melted rock. The northern side, being probably the weakest, gave way, and then the lava burst out into the plain below. Taking at once an easterly course, owing to the general slope of the ground, it descended in a sheet of dark rugged rock, now swelling up against ridges that opposed its progress, and then sweeping past them until it reached the beginning of the hill of Pradelle already noticed. Here, in a scene of singular confusion, it broke into two streams, one leaping like a torrent down the valley of Villar, the other plunging into the valley of Gresinier.[1] But the emission of this vast body of melted rock did not conclude the eruptions of the Puy de Pariou. When the lava had perhaps ceased to flow, the vapour and gases still continued to escape with violence. By their means another cone was in time produced, not quite on the former site, but, as so often happens, a little to one side, so as to cover the southern half of the older cone, and leave visible that northern segment of it from which the lava issued. Thus arose the later cone of Pariou. No subsequent eruptions have disturbed its regularity or filled up its crater. The hand of time has not effaced its smooth curves and slopes, but has covered them with vegetation, whereby the loose dust and scoriæ are protected from the destructive effects of heavy rains. After the lapse

[1] Mr. Scrope's description of these lava-streams is a model of graphic and accurate description.

Fig. 6. — View from the top of the Puy de Pariou.

of many a long century this little volcano is still nearly as perfect as when the last shower of ashes fell over its sides, and it promises to remain so for centuries to come.

The Puy de Pariou is only one of a series of similar cones. Some have but one crater, others have two, three, or even, as in the instance already cited, four. Each crater is of course the product of a different eruption or series of eruptions, as the Puy de Pariou so well explains. Several striking examples of the bursting of the side of a cone by the pressure of the uprising column of lava within it, occur among the cones to the south of the Puy de Dôme, as in the Puy de las Solas and the Puy de la Vache. These two hills, when seen from the south, look like the mouths of two yawning chasms. Their southern sides have been swept away by a black rugged river of lava, which, issuing from the bottom of each crater, flows eastward in a united stream for twelve miles down a deep, narrow valley. The scenery round these hills is even more desolate than among those to the north of the Puy de Dôme. The cones and craters are in many places devoid of all verdure, and have still much of the blackened and burnt aspect of active volcanoes. The lava, too, which has spread out over most of the intermediate ground, is dark, bristling, and sterile. The whole landscape leaves an impression, not easily effaced, of the vigour of volcanic agency, and of its power to modify, and even altogether change the general aspect of a district.

To one who had been at work for some years among a set of old and fragmentary volcanic rocks, trying to piece together porphyrites, dolerites, basalts, and tuffs, the sight of those Puys, with their fresh cones and craters of ashes and scoriæ, and their still perfect floods of lava, was inexpressibly instructive. Merely to cast the eye over the landscape was of itself a memorable lesson. The scene was exactly what was needed to enable one to realise the character of those old British Carboniferous volcanoes of which only such mere fragments now remain. High among the uplands of Central France my eye was ever instinctively recalling the hills and valleys of Central Scotland, and picturing their original scenery by transferring to them some of the main features in the landscapes of Auvergne. The imagination easily filled again with a sheet of deep blue water the broad expanse of yonder Limagne. Vines, and acacias, and mulberry-trees, seemed to melt of their own accord into stately *sigillariæ, lepidodendra*, and *calamites*; the orchards and cornfields along the slopes began to wave with a dense underwood of ferns and shrubby vegetation; some of the cones rose fresh and bare, others were dark with a growth of araucarian conifers, and there, with but little further change, lay a landscape in the central valley of Scotland during an early part of the great Carboniferous Period. Nor did a more extended examination of other parts of the Volcanic District weaken this comparison, for the general outward resemblance of the present volcanic rocks of France, to what must have been the original aspect of those of Scotland at the geological era just named, holds good, even when traced into detail.

One of the most interesting excursions from Clermont is to the hill of Gergovia, about six miles to the south. We started off early one morning, while the sky, which had been remarkably clear for some days, began to grow dusky with heavy clouds that kept trooping up from the south west. Puy de Dôme had his head wrapped in mist, and giant shadows chased each other across the range of Puys until, as the clouds thickened, all the uplands were shrouded in an ominous gloom. Rain at last began to fall in large round drops, and a distant muttering of thunder was heard rolling away northward. But the morning being fresh and cool, even at the risk of a good drenching we persevered. The road, like all the French military highways, excellently made and well kept, passes through endless vineyards, many of which lie among the broken ruins of lava-flows that have descended from the heights to the westward. At one point it has even been cut through a part of one of these lavas.

The hill of Gergovia is famous in history as the site of a town long and successfully defended by the Arverni (people of Auvergne) against Cæsar's legions. Some interesting antiquarian remains had been found shortly before our visit, and we learnt that excavations were about to be renewed in search of more. But the hill is not less interesting to the geologist than to the antiquary. Seen from the east, it looks like a broad truncated cone; but it differs altogether in appearance and origin from the true volcanic cones of the Puys. It consists, in fact, of horizontal strata of marl and limestone; about two-thirds of the way up lies a bed of basalt, which forms a marked feature along the hillside; some calcareous and ashy strata next occur, while the summit is formed by a capping of basalt. These marls and limestones are of lacustrine origin, as is shown by their fresh-water shells, and by the caddis-worm cases which they contain. Forming parts of the deposits of the old lake of the Limagne, they attain in this hill a thickness of probably not less than 1200 or 1500 feet. Ascending one of the ravines which deeply furrow the east side of the hill, we passed over these thinly-laminated strata, piled over each other in successive layers, and crumbling away like chalk. Every yard of the steep ascent deepened the impression of the exceedingly slow rate at which these sediments must have been formed, and therefore of the prodigious lapse of time which their entire thickness represents. The morning, after clearing up for a brief space, had again overcast, and rain began to fall as heavily as before. We sheltered for a little under the lower basalt. Had we been suddenly spirited away unawares from some of the Scottish glens, and set down at the side of this rock, we should hardly have recognised the change of scene. The basalt is a true bed, some thirty or forty feet thick, and is scarcely distinguishable from certain Carboniferous basalts of the Lothians. It is a hard, dark, compact rock, somewhat rough and scoriaceous towards the bottom, like the basalts along the magnificent coast-section near Kinghorn in Fife. But what especially interested me was, to find that the upper surface of the bed was even and smooth, and that the marls rested on it unaltered, the line of demarcation being sharp and clear. The basalt had undoubtedly rolled over the bottom of the old lake; it rested on lacustrine marls, and strata of the same kind covered it. But its upper surface, so far from rising up into black bristling masses, like the subaerial currents of the Puys, was smooth and even, like the top of a bed of sandstone or limestone, and the marls which succeeded gave no sign of alteration or disturbance. I therefore inferred that the evenness of the upper surface of many Palæozoic and Tertiary basalts in Scotland offered no valid objection to their being of the nature of true lava-currents, poured out at the surface, and not injected at some depth beneath it.

Ascending beyond the prominent zone of basalt, we soon reached a bed of calcareous peperino, or tuff, that at once recalled some of the tuffs associated with parts of the Carboniferous Limestone of Linlithgowshire and Fife, its stratification being confused, sometimes highly inclined, changing its direction, or even disappearing altogether. Similar ashy materials, mingled with calcareous matter, occupy the remainder of the hill up to the cake of basalt which crowns the summit, and show how among the fine sediments of the ancient lake volcanic ejections were occasionally thrown down.

We intended to make a circuit of Gergovia, descending on the north-west side towards the strange isolated castle-crowned crag of Montrognon. But the rain, which had fallen with scarcely an intermission since we began the ascent, now came down in torrents. We took refuge in a little cave in the calcareous peperino, which looked eastward across the Limagne to the distant mountains of the Loire and southward to the volcanic heights of the Velay. But the landscape was blotted out in so thick a veil of falling water that we could hardly distinguish the form of the trees at a short distance down the slopes. It was an instructive lesson in denudation to sit at the mouth of the cave and watch the increase of the runnels. Over ground which in the morning was as dry and parched as a drought of some weeks' duration could make it, water now poured in hundreds of rivulets, acquiring a milky colour from the marl *débris* which it swept away in its descent. One could see how rapid must be the waste of these soft calcareous rocks. Baked and cracked by the fierce heat of summer, their surface crumbles down. Every shower loosens and removes portions of this disintegrated surface and prepares the way for the action of the shower that succeeds. It is by these means, joined with the undermining agency of rivers, that the deep and wide valleys of these districts have been excavated.

Sitting in the cave while the deluge continued outside, we had leisure to reflect on the geological history of the hill. Its strata were elaborated at the bottom of the lake that filled the broad valley of the Limagne. Leaf after leaf, and layer after layer of marl and limestone were slowly laid down, derived mainly from the crumbling remains of shells, cyprids, and other living creatures that tenanted the water. The rate of growth of these tranquil deposits must have been remarkably slow. When a thickness of at least a thousand feet of them had been formed, a volcano sprang up in the neighbourhood, and rolled into the lake the stream of lava represented by the lower bed of basalt. Fine calcareous sediment, however, began to be deposited anew over the floor of lava, yet the volcanic forces had not become wholly quiescent, for from time to time showers of ashes were thrown out, which, falling into the lake, gave rise to those beds of peperino, in one of which we were now taking refuge from the storm. Afterwards another stream of lava was erupted, forming the present summit of the hill. How much farther the series may have originally extended cannot now be discovered, since if anything was deposited on the surface of the second basalt it has been subsequently worn away. The rain at last ceasing, we descended by an endless series of turnings and windings to a tree-shaded road that led through cornfields, now heavy with their golden crop. Away to the left we could see the Château de Montrognon, a ruined fortalice perched on the summit of a narrow and precipitous basaltic hill. Farther over lay the high ground of the Puys, with the rain-clouds still floating over it. As we advanced, however, the sky began to clear, patches of deep blue now and then appeared through gaps in the driving clouds, until the last mist-wreath rose from the great Puy de Dôme, and amid gleams of bright sunshine we re-entered Clermont about noon.

The journey to Mont Dore, being uphill nearly all the way, takes the greater part of a day. The first half of the road winding up the side of the granitic plateau crosses several of the lava-streams which have descended the valleys, like that from the cone of Pariou, and at last reaches the desolate tableland on which runs the chain of the Puys. A good view is obtained of several of the cones on the south side of the Puy de Dôme, the ruined yawning craters of the Puy de las Solas and the Puy de la Vache being especially noticeable, with their now silent rivers of black rugged lava. From the half-way house the road runs southward over the undulating surface of the plateau, until it begins the ascent of the Mont Dore hills. These heights, in their lower portions, are tolerably green, and constantly recall to my memory parts of the basaltic scenery of Skye and Mull. Numerous blocks of basalt, sometimes of considerable size, are scattered over the surface, and often lie in such positions that it is difficult to see how the action of the atmosphere, or of running water, could have placed them there. I kept an eye on the alert to detect a striated or polished surface; but there is little rock exposed in places along the road, and I was unsuccessful. It seemed at the time, however, to be far from unlikely that some of these great blocks of stone had been ice-borne. When the glaciers of the Alps filled the valley of the Lake of Geneva, at a height of no more than 1200 feet above the sea, there seems no reason why glaciers should not have descended from the Mont Dore mountains, which now form the highest ground in Central France, rising in the Pic de Sancy to a height of 6217 feet. At this day, indeed, snow remains unmelted in the higher recesses of these mountains even in midsummer. I am not aware, however, that the existence of glaciers has ever been recognised here, and I had no time even to make any attempt to solve the question for myself. The occurrence of the scattered blocks, and of some coarse unstratified detritus, in the steep defile that descends from the east into the valley of the Dordogne, was at least sufficient to suggest the possibility of a partially glacial origin for some of the deep valleys of the Mont Dore.[1]

The Baths lie in a valley of surpassing loveliness, hemmed in by lofty mountains and huge precipices. The climate is delicious as a contrast to the scorching sultriness of the lower plains, and hence the locality has been a watering-place since the days of the Roman occupation of Gaul. We had time only to get a peep at the conglomerates and trachytes of this great volcanic district. Everything is on a scale so much vaster than in the country of the Puy de Dôme, that the first impression of the geologist is one of bewilderment. We did not remain long enough to get rid of this feeling, and at this moment I have a confused remembrance of vast irregular sheets of trachytic lava, separated by piles of volcanic ash and conglomerate, the whole thrown together in a way which at the time it seemed hopeless to attempt to unravel; of dykes and veins of basalt, and currents of lava, belonging to much more recent eruptions that flowed down the deep valleys which had been excavated out of the ancient lavas.

Contenting ourselves with a mere survey of its external features, we left the Mont Dore district by the road which, on re-ascending from the valley of the Dordogne, strikes towards the east and then sweeps down into the valley of Chambon. The Baths, after lying for some hours under the shade of the great hills, were bathed in sunlight, and full of bustle, as we drove through the streets. Invalids, valetudinarians, and fashionable visitors, may be seen passing

[1] Since this essay was published the former existence of glaciers in Auvergne has been shown by MM. Delanoue, Marcou, and Gruner.

to and fro between the hotels and the central building where the waters are dispensed. Some are borne in sedan-chairs, but the greater number preform the short journey on foot. Men and women, as soon as they imbibe their draught, hurry home holding their mouths—a sight which is now and then irresistibly comic—as where a portly priest, perhaps of some threescore, shuffles back to his hotel with the ends of his dress muffled round his mouth and nose. On inquiry we learned that this proceeding is meant to prevent the gas from escaping after the morning dose of water—a precaution without which it is held impossible to derive the full benefits of *les eaux minérales*.

The journey from Mont Dore les Bains to the plain of the Allier at Issoire is probably one of the most interesting in Central France. From the summit level of the road the eye wanders over a wide sweep of mountains of volcanic origin, traversed by wide valleys and narrow gorges. Southward, in the dark shady rifts of the higher peaks, lie gleaming patches of snow, and the breeze that plays about these uplands, even in the bright sunshine, is cool and refreshing. In the course of the descent we again observed evidence of lava-flows of several distinct ages, some of them high up along the sides of valleys which had since been excavated through them; old river gravels, too, far above the channels of the present streams; and in the bottom of the valley, following all its curves like a river, a current of black rugged lava, which in one or two places rose up into the most fantastic masses. The impression of the immense lapse of time represented by these Tertiary formations and their subsequent denudation was deepened tenfold as we threaded this valley of Chambon. The stream which meanders through the broader meadow-lands, and leaps down the narrower defiles, has undoubtedly been the main agent in scooping out this great indentation in the flanks of Mont Dore. Here and there, in the centre of the valley, it has left isolated patches of the beds of rock that occur on either side, such as the picturesque conical crag on which stands the ruinous castle of Murol. These outliers are silent witnesses of the reality of the erosion. The lava-current at the bottom of the valley has certainly not been erupted since the time of the Romans. It must, therefore, be at least 2000 years old, and may, for aught we can tell, be ten or a hundred times older. Yet since its eruption, the action of the river, though here and there bisecting the lava, has nevertheless been, on the whole, but trifling; indeed the amount of excavation effected since the eruption of this lava probably falls far short of a thousandth part of the general erosion of the valley. Yet the excavation of the valley of Chambon is the latest and perhaps the shortest of all the stages which the geology of the district indicates. How vast must have been that earlier period wherein were deposited those fine alternations of lime and clay which form hills, such as Mont Perrier, several hundred feet in height, divisible into distinct zones, each characterised by peculiar assemblages of fossils. It is only by thus advancing, step by step, backward into the remote past, that we begin to appreciate the antiquity of the Tertiary groups of strata, and to realise, in some measure, the extent of that long history of physical and organic change of which these strata contain only the last chapters.

We hurried onward from Issoire up the plain of the Allier, catching a glimpse of the little contorted coal-field of Brassac—an outlier of true Carboniferous strata, resting in a hollow of the crystalline schists, and overlapped by Tertiary marls and limestones which stretch southward from the Limagne. Here and there in the valley were volcanic mounds, sometimes capped with little towns, so that, although we had quitted the district of great lava-streams, we were far from having reached the limits of the volcanic district. The town of Brioude lies at the southern extremity of that great lacustrine deposit of the valley of the Allier, so conspicuously displayed in the Limagne d'Auvergne. The granitic hills close in upon the river, and thence swell southward into the mountains of La Margeride and the uplands of the Haute-Loire. Of Brioude itself I have a pleasant recollection as a quaint rambling town with some large decayed houses that seem to have once been tenanted by a better class of inmates. The hotel at which we stayed was one of these. From a retired street we entered a low archway, and found ourselves in a dark room with a large fireplace, now used as a kitchen. A number of doors opened out of the farther side of the room, and through one of them we were ushered into a lobby with broad staircase and carved banisters. Up and down, through one passage into another, we at last halted at a recess on one of the landings, and were shown into a large wainscoted bedroom. Its tarnished mirrors, faded green-velvet chairs, old-fashioned cabinets and tables, were certainly not the kind of furniture one would have expected to see in a quiet hotel in a remote little town. There was a taste and harmony about the whole, and they fitted so well with the character of the rest of the house, as to suggest that the place had been the residence of some decayed family, and that not many years could have elapsed since it passed into the hands of an innkeeper.

Crossing the Allier by the fine bridge at Old Brioude, and bidding adieu to that noble river, we started for Le Puy. Our course lay towards the south-east, up a range of granitic heights, traversed by numerous narrow and deep, but often thickly-wooded ravines, and with fragments of ancient basalt now and then protruding by the roadside, or along the upper edge of a steep bank. The country, however, remains somewhat bare and uninteresting; nor until one begins to descend towards the basin of the Loire, and catches sight of the range of volcanic hills and cones that encircle Le Puy, does its interest revive.

Le Puy is one of the most picturesque towns in France, built round a conical hill, which rises in the valley between the River Borne and another tributary of the Loire. An abrupt crag of breccia, crowned with a bronze statue of the Virgin, overhangs it on the north; while lower down in the plain a tall massive column of the same rock supports the small and seemingly inaccessible church of St. Michel. The country rises rapidly on all sides, so that Le Puy lies embosomed among hills—vast piles of lava, and cones of ash formed by many different eruptions, sweeping away south into the heights of Mont Mezen and the long plateau which here separates the waters of the Allier from those of the Loire.

The geologist could hardly pitch upon a locality where more may be learned in so narrow a compass. Le Puy lies in the centre of another Tertiary lake, some twenty miles long, and twelve or fourteen broad. This lake occupied a hollow in the great granitic framework of the country, and, like the Limagne d'Auvergne, gave rise to the slow accumulation of fine marls, limestones, and sandstones, which attained a united thickness of hundreds of feet. Over the top of these horizontal strata, lavas and ashes were erupted to a depth of three or four hundred feet, so as wholly to cover up the lacustrine deposits, and obliterate the site of the lake. Since these events, the Loire and its tributaries have been ceaselessly at work in deepening and widening their channels. And now, incredible as it may seem, these streams have actually cut their way down through the solid basalt, and a great part of the old lake formations. They have, in short, excavated a series of valleys, several hundred feet deep, and sometimes of considerable width, along the sides of which are exposed the remaining edges of the strata that have been worn away. Standing on the summit of the Montagne de Denise, and looking round upon the valleys and ravines on every side, each traversed by what seemed such an insignificant stream, I felt as if a new geological agent were for the first time made known to me. Striking as are the proofs of erosion in the country of the Limagne, they fall far short of these in the Haute-Loire. To be actually realised, such a scene must be visited in person. No amount of verbal description, not even the most careful drawings, will convey a full sense of the magnitude of the changes to one who is acquainted only with the rivers of a glaciated country such as Britain. The first impression received from a landscape like that round Le Puy is rather one of utter bewilderment. The upsetting of all one's previous estimates of the power of rain and rivers is sudden and complete. It is not without an effort, and after having analysed the scene, feature by feature, that the geologist can take it all in. But when he has done so, his views of the effects of subaerial disintegration become permanently altered, and he quits the district with a rooted conviction that there is almost no amount of waste and erosion of the solid frame-work of the land which may not be brought about in time by the combined influence of springs, frost, rain, and rivers.

The volcanic phenomena of the neighbourhood of Le Puy are likewise full of interest, and, owing to the numerous deep ravines, they can be easily studied in admirable natural sections. The sheets of lava, often beautifully columnar, recall many of the basalts of Scotland. The beds of peperino, or tuff, likewise bear the strongest resemblance to some of the Carboniferous tuffs of the Lothians. Indeed, many parts of the scenery differ but little from some of the Scottish volcanic districts. We found the cones of scoriæ more numerous, but less perfect than round the Puy de Dôme; as if they belonged to an earlier era, and had consequently been longer exposed to the wasting effects of time. But this greater antiquity is occasionally productive of much advantage to the geologist, for it presents him with chasms and cliffs, without which he would miss many incidents in the geological history of the district. Thus, near Le Puy, the volcanic cone of Mont Denise, so well known for the interesting fossils which have been found in its underlying gravels, has had its western front exposed partly by nature and partly by man. By this means are laid bare the strata of volcanic breccia that rest on the marls of the old lake; on a worn surface of the breccia comes a band of true river gravel now several hundred feet above the present bed of the Borne, while associated with this gravel there is sometimes a newer volcanic tuff. Through these various deposits the volcano of Mont Denise broke out, piling up the mound of loose scoriæ and ashes that form the hill. Here we saw, what it had not been our good fortune to meet with in the Puy de Dôme—the actual section of a volcanic vent. The sides were smooth and worn, and the bed of hard breccia, which had been perforated nearly vertically, still retained the grooving and polishing produced by the friction of the ejected scoriæ. The vent was filled up with a black scoriaceous lava, while several lava *coulées* that had rolled down the hillside now formed dark masses of prominent crag and cliff. This little volcano bore a close resemblance to the upper part of Arthur's Seat at Edinburgh. In each case a column of lava is surrounded by an outer envelope of loose ashes, over which various currents of lava have rolled down from the crater.

With no little reluctance, and not until the sun had dipped behind the western hills, did we quit the slopes of Mont Denise. The evening, after a day of mingled storm and sunshine, was beautiful, and the whole of that wondrous landscape lay bright and clear around. It was the last evening, too, which we had to spend in the volcanic region of Central France; nor could we have secured a more auspicious sky or a more favourable locality for taking a last view of the scenery and summing up the results of the journey. Sitting on a pile of loose cinders on the top of the hill, we watched the level rays lighting up the vast basalt plateau that stretched away for miles to the west, while each of the many cones that dotted the plain cast its long shadow towards us. With undiminished wonder we gazed again at the deep ravines and valleys by which the plateau is broken up, each with its streamlet meandering like a silver thread between the slopes. The sunlight lay warm and bright on the town of Le Puy in the valley below, with its isolated crag of La Vièrge, and its church-crowned pinnacle of St. Michel—two rocks that remain to record the enormous erosion of these valleys. The castle of Polignac—built on another outlying crag farther down the plain—stood up in the deep shadow of Mont Denise. Eastward, the gorges that open into the Loire gleamed white as the sunset fell along their bars of pale marls and limestones, and their capping of basalt. Beyond these, cone rose behind cone, amid piles of lava-currents of many different ages; each sunward slope and crest was now flushed with a rosy hue deepening into purple in the distance, until, far away as the eye could reach, the mountains of Mont Mezen were steeped in the softest violet, that melted into the twilight of the eastern sky.

And here we took leave of the volcanoes of Central France. Coming as learners to a district which had been already often and carefully explored, we gained such a vivid impression of the phenomena of the country as can only be obtained from an actual visit. We were now able to realise, with a clearness till then unlooked for, the original features of those ancient Scottish igneous rocks, among whose fragmentary relics we had been at work for years. In the form of their cones, their distribution, their aspect in the landscape, the limited extension of their ashes, the form and disposition of their lava-currents, the structure of their craters, and their relation to the underlying and to the contemporaneous stratified deposits, these extinct Tertiary volcanoes of France cast a flood of what to me was new light upon the long-extinct Carboniferous volcanoes of Scotland. I seemed no longer to be dealing with conjectures, but with sober truths. To the history of the igneous rocks of my own country there was now imparted a freshness and reality such as it did not possess before. More than ever did these rocks stand forth, not as mere mineral masses, to be described in text-books as occupying definite areas of ground, or to be arranged by hand-specimens in a museum as so many mineralogical compounds, but as the records of a long geological history which they would unfold if only questioned in the right way. And the main result of our wanderings in the Auvergne and Velay was to show us how this questioning should be carried on.

Nor did we value less the new and enlarged views which those rambles gave us of the potency of rain, rivers, and other atmospheric agencies, in effecting the degradation of the land. Nothing we had read in geological literature, not even Mr. Scrope's classic descriptions of this very region, had prepared us for the contemplation of changes so stupendous as those of the erosion of the ravines and valleys of Le Puy. To look upon them for the first time was, as I have said, like a new revelation, which in an instant uprooted a host of narrow long-cherished conceptions, and supplanted them with a profound respect for the power of the terrestrial agencies of waste. Broader, and truer, and fresher views of nature are worth the trouble of a long journey, and in gaining them we felt ourselves abundantly repaid for our toil under a fierce sun among the uplands of Central France.

VI.

THE OLD GLACIERS OF NORWAY AND SCOTLAND.[1]

IN the course of the detailed investigations of the history of the glacial period in Britain, which, during the past six or seven years, have been carried on by the Geological Survey, the desire naturally arose to compare the phenomena of glaciation now familiar in this country with those of some other region where they might be linked with the action of still existing glaciers. No other part of Europe offered so many facilities for such a comparison as were to be found in Scandinavia. In the first place, the rocks of the two regions were known to present many points of resemblance in structure and scenery. It was further evident from the published accounts that the Norwegian coast possessed the ice-worn aspect so characteristic of the West of Scotland.

The objects proposed to be accomplished in this excursion were—to compare, as minutely as time would allow, the ice-marks on the rocks of Scotland with those

---

[1] *Proc. Roy. Soc. Edin.*, January 1866. The observations here recorded were made by me in the summer of the year 1865, in company with my colleagues in the Geological Survey, Messrs. W. Whitaker and James Geikie.

on the rocks of Scandinavia; to ascertain, from personal exploration, how far the glaciation of the Norwegian coasts and fjords could be traced to the action of land-ice or of floating bergs; to trace, if possible, the connection between the ancient ice-work and the work of living glaciers; and, generally, to seek for any facts that might help to throw light upon the history of the glacial period in the British Isles. Having only a few weeks at our disposal, we were far from aiming at original discovery in Norwegian geology. The main features of the disposition of the snow-fields and glaciers had already been given in the masterly sketch by Principal Forbes — a work which was of inestimable value to us.[1] More detailed descriptions of parts of the glaciation of Norway had been published by Scandinavian geologists—Esmark,[2] Hörbye,[3]

[1] *Norway and its Glaciers*, 8vo, 1853. Mr. Chambers also has referred to the striated rocks in different parts of Norway in his *Tracings of the North of Europe*, 1850.

[2] Esmark. *Edin. New Phil. Journal*, vol. ii. p. 116 *et seq.* (1826). In this paper the former presence of land ice over large areas from which it is now absent, and its powerful influence as a geological agent of abrasion, are, for the first time, distinctly recognised. The illustrations are taken from the south of Norway.

[3] Hörbye. "Observations sur les phénomènes d'erosion en Norvége"—*Programme de l'Université de Christiania pour* 1857. The author gives a careful *résumé* of all the observations made by himself and others upon the direction of the striæ on the rocks of Norway, and adds a number of maps, one of which shows the outward radiation of the striæ from the central mountain mass of Scandinavia. Yet he commits himself to no theory as to the nature of the agent by which the striæ were produced. In a concluding section upon the glacial theory, he says :— "Il est vrai sans doute qu'en général la direction des stries est parallèle à l'avancement des glaciers actuels ; mais je ne vois pas que cette circonstance puisse suffisamment démontrer que les stries ont été gravées par les glaciers." "Je me joins à cette conclusion, que les sulcatures du Nord se présentent comme des

Kjerulf,[1] Sexe,[2] and others. Yet I was not without the hope that, besides adding to our own experience, we might also be fortunate enough to find in the Norwegian fjords materials for making still more clear the geological history of our own western sea-lochs.

The close resemblance between the general outline of Scotland and that of Scandinavia is too well known to need more than a passing allusion. The numerous deep and intricate indentations, the endless islands and skerries, the mountainous shores, the host of short independent streams on the western coast; and on the eastern side, the broad, undulating lowlands, sending their collected drainage into large rivers, which enter the sea along a comparatively little embayed coast-line, are familiar features on the maps of both countries. This general outward resemblance, which at once arrests the attention of every traveller in Norway to whom the scenery of the Western Highlands is familiar, depends upon a close similarity in the geological structure of the rocks, and a coincidence in the geological history of the surface of the two regions. Norway, from south to north, is almost wholly made up of crystalline and schistose rocks, not all of the same age, yet possessing a general similarity of character. In like manner, the West of Scotland, from the Mull of Cantyre to Cape Wrath, is in great measure

produits qu'un agent plus puissant et plus général que les glaciers dont l'action conserve toujours un caractère plus local." But he does not indicate what this more powerful and more general agent may be.

[1] Kjerulf. *Uber das Friktions-Phænomen*, Christiania, 8vo, 1860. See also *Programme de l'Université de Christiania pour* 1860, and *Zeitschrift der Deutsch. Geol. Gesellschaft*, 1863, p. 619, and plate xvii.

[2] Sexe. "Om Sneebræen Folgefon." Christiania. *Universitets-program for andet Halvaar* 1864. This paper gives a detailed account, with map and sections, of the Folgefon snow-field and its glaciers, including the well-known glacier of Bondhuus.

built up of gneisses, schists, slates, quartzites, granites, and other rocks, quite comparable with those of Norway.[1]

Besides the external resemblance due to the lithological nature of the rocks beneath, there is a still further likeness dependent upon similarity, partly of geological structure, and partly of denudation. Most of the Scottish sea-lochs have had their trend determined by lines of strike or of anticlinal axis, and the same result seems to have taken place in Norway. But the lochs and glens of the one country, and the fjords and valleys of the other, whether or not their site and direction have been determined by geological structure, unquestionably owe their excavation to the great process of denudation which has brought the surface of the land to its present form.[2] In short, Norway and the Scottish Highlands seem to be but parts of one long tableland of erosion composed of palæozoic (chiefly metamorphic) rocks. This tableland must be of venerable antiquity; for it seems to have been in existence, at least in part, as far back as the Lower Old Red Sandstone. Since that time it has been sorely defaced by long cycles of geological revolution; rains, rivers, ice, and general atmospheric waste have carved out of it the present valleys, and to all this surface-change must be added the results of dislocations, as well as unequal upheavals and depressions of the crust of the earth beneath. Nevertheless it still survives in extensive fragments in Norway, where it serves

[1] Since this paper was published, my friends Dr. T. Kjerulf and Dr. Tellef Dahll have given to the world numerous instructive memoirs on Norwegian geology. A German translation of Dr. Kjerulf's *Geology of Southern Norway* has been published by Dr. Gurlt, Bonn, 1879.

[2] I have tried to trace the history of this process in the case of the Scottish Highlands. *The Scenery of Scotland viewed in connection with its Physical Geology*, chap. vi.

VI] OLD GLACIERS OF NORWAY AND SCOTLAND. 113

as a platform for the great snow-fields, while it can even yet be traced along the undulating summits of the mountains of the Scottish Highlands. One of its latest great revolutions was a submergence towards the west, which, extending from the coasts of Ireland to the north of Norway, has given rise to some of the most distinctive features

Fig. 7.—Ice-worn bosses of gneiss and perched blocks. North coast of Sutherland.

of that part of Europe. No one can attentively compare the maps of the land with the charts of the sea-bottom in the region between the headlands of Connaught and the North Cape, without being convinced that the endless ramifying sea-lochs and fjords, kyles and sounds, were once land-valleys. Each loch and fjord is the submerged part of a valley, of which we still see the upper portion above water, and the sunken rocks and skerries, islets and islands are all so many relics of the uneven surface of the old land before its submergence. The indented form of the coast-line of the west of Scotland and of Norway is not evidence of the unequal encroachment of the sea, as is often, perhaps generally, supposed, but is due to a general submergence of the west side of the two countries, whereby the tides have been sent far inland, filling from side to side ancient valleys and lakes.[1] Subsequent re-elevations, or rather, stationary intervals during a long period of elevation, are marked along both the Norwegian and Scottish shores by successive terraces or raised beaches.

But to one who has sailed or boated among the sea-lochs of Scotland, no feature of the Norwegian coast is at once so striking and so familiar as the universal smoothing and rounding of the rocks, which is now recognised as the result of the abrading power of ice. Every skerry and islet among the countless thousands of that coast-line is either one smooth boss of rock, like the back of a whale or dolphin, or a succession of such bosses rising and sinking in gentle undulations into each other. Such, too, is the nature of the rocky shore of every fjord; the smoothed surface growing gradually rougher, indeed, as we trace it upward from the sea-level, yet continuing to show itself, until at a height of many hundred feet it merges into the broken, scarped outlines of the higher mountain-sides and summits.[2]

---

[1] See a fuller statement of this subject in *Scenery of Scotland*, pp. 125-137.

[2] The singularly ice-worn aspect of the Norwegian coast, as well as its strong resemblance to the west coast of Scotland, was succinctly described by Principal Forbes, *Norway and its Glaciers*, p. 42 *et seq.*

VI] OLD GLACIERS OF NORWAY AND SCOTLAND. 115

In short, as is now well known, the whole of the surface of the country, for many hundred feet above the sea, has been ground down and smoothed by ice.

We sailed along the coast of Norway, between Bergen and Hammerfest, by the usual steamboat route, touching at many stations by the way, threading the narrow kyles and sounds that lie among the innumerable islands, and now and then running inland up some fjord far into the heart of the country. We halted here and there to spend a few days at a time in exploring some of the fjords and glaciers. What can be seen from the steamer on the coasting voyage is now familiar from the numerous descriptions which have been given of it in recent years. I shall therefore content myself with offering an account of two excursions to points at some distance from the ordinary route.

A little to the north of the Arctic Circle lies the island of Melö, one of many which are here crowded together along the coast. It is only noticeable, inasmuch as it is a station at which the steamers call, and from which the great snow-fields of the Svartisen or Fondalen may be most easily visited. Here, as along all the Norwegian coasts, we find ourselves among bare bossy hummocks of rock thoroughly ice-worn. From the higher eminences the eye sweeps over the countless islets and skerries, and far across the Vest Fjord to the serrated peaks of the Lofodden Islands, which in the distance seem deep sunk in the north-western sea. The whole of the lower grounds is one labyrinth of *roches moutonnées*, raising their smooth backs like so many porpoises out of the sea, as well as peering out of a flat expanse of green pasture and dark bog which here covers an old sea-bottom. The striations and groovings are still fresh on many of the smoothed surfaces of gneiss, and invariably run straight out to sea in the line of the long valley up which the sea winds inland among the snowy mountains. It cannot be doubted that a vast mass of ice has come seawards down this valley, and that all these ice-worn hummocks of rock were ground down by it. The wide opening at Melö is formed by the converging mouths of a number of narrow fjords (Fig. 8). Of these the most northerly is the Glommens Fjord, which is bounded along its northern side by a range of high mountains, with a serrated crest and abundant snowy clefts and corries. Southward lies a belt of lower ice-worn hills, cut lengthwise by the Bjerangs Fjord, and bounded on the south by the Holands Fjord, on the south side of which rises another range of scarped snow-covered mountains.[1]

From the *gaard* of Melö we boated eastward among various small islets and channels, passing soon into the Holands Fjord, up which we continued until we rested underneath the great snow-field and glaciers of Svartisen. In this excursion we started from the coast, amid islands, all moulded, like those of the West of Scotland, by the ice of the glacial period, and in the evening we reached rocks on which the present glaciers are inscribing precisely the same markings. One of the first features which arrested attention was the contrast between the smoothed, ice-worn surface of the lower grounds and the craggy, scarped outlines of the mountain crests. This was especially marked along the northern side of the Glommens Fjord, where the ice-worn rocks form a distinct zone along the side of the

[1] Although I use the word *mountains*, there is no definite system of ridges; on the contrary, these fjords must be regarded as indentations along the edge of a great tableland, of which the average level may range from 3000 to 4000 feet above the sea, and which serves as the platform on which the wide snow-fields lie. See *Norway and its Glaciers*, pp. 190 232.

rough, craggy hills. To the north of Melövaer this ice-worn belt was estimated to rise about 200 feet above the sea. Its smoothed rocks are abundantly rent along lines of joint and other divisional planes; their ice-worn aspect must thus be imperceptibly fading away. The rough rocks above them sometimes show traces of smoothed surfaces, as if they too had suffered from an older glaciation, of

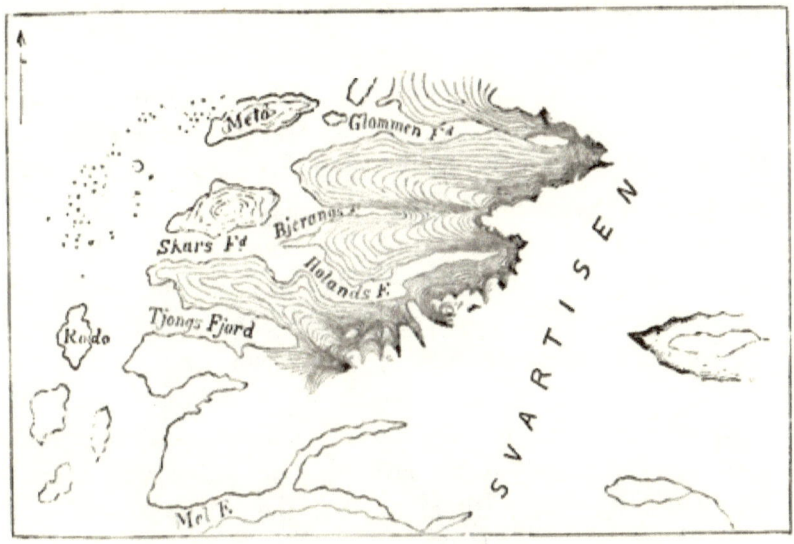

Fig. 8.—Map of the Neighbourhood of the Holands Fjord (Munch).

which the records are now all but obliterated. The line of division between the belt of rocks which have been smoothed by ice, and those which have been roughened and scarped by atmospheric waste, slopes gently upward in the direction of the central snow-fields of the interior. While at Melövaer it seemed to rise only about 200 feet above the sea; at Fondalen, twenty-five or thirty miles inland, it mounts to a height of fully 1500 feet. A tract of bare hills, lying between the Glommens and the Holands Fjords, and rising eastward into the snow-covered tableland, is well smoothed in the direction of these fjords. In short, the whole of the broad depression between the two fjords has been filled with ice, moving steadily downwards from the snow-fields to the sea.

It was interesting to watch, on every little islet and promontory under which we passed, even the same details of glaciation so familiar along the margin of our Scottish fjords. The rocks, smoothed into flowing lines, slip sharply and cleanly into the water, and are well grooved and striated. Moreover, it was easy to see that the ice which had graven these lines must have moved down the fjord, for the *lee* or rougher side of the crags looks seawards. It was likewise clear that the scorings were not the work of drifting bergs or coast ice, for they could often be seen mounting over projecting parts of the banks, yet retaining all the while their sharpness, parallelism, and persistent trend. Another point of similarity to West Highland scenery was found in the strange scarcity or absence of drift and boulders. I do not mean to assert that these are not to be met with at all, but they do not exist so prominently as to catch the eye even of one who is on the outlook for them. The rock everywhere raises its bare knolls to the sun as it does on the coasts of Inverness and Argyll. To complete the resemblance, the Norwegian fjord has its sides marked by the line of a former sea-margin, about 250 feet above the present. This terrace winds out and in among all the ramifications and curves of the fjord, remaining fresher and more distinct than the raised beaches of the West Highlands usually are, and even rivalling one of the parallel roads of Lochaber.

We rested for a week at the hamlet of Fondalen, on

Fig. 9.—View of the two Glaciers of Fondalen, Holands Fjord.

the south side of the Holands Fjord. It stands at the mouth of a deep narrow valley on the line of the terrace, which here runs along the crest of a steep bank of rubbish covered with enormous blocks of rock—an old moraine thrown across the end of the valley. There seems to have been at one time a lake behind this bank, formed by the ponding back of the drainage of the valley, and gradually emptied as the outflow-stream deepened its channel through the moraine. At the head of the valley a small glacier descends from the snow-field of Svartisen. There could be no better locality for studying the gradual diminution of the glaciers, and for learning that it was land-ice that filled the Norwegian fjords, overrode the lower hills, and went out boldly into the Atlantic and Arctic Sea. The Holands Fjord runs, as I have said, approximately east and west, and this short narrow valley descends from the south. The fjord was formerly filled with ice, and is therefore polished and striated along the line of its main trend. The valley of Fondalen was likewise filled with ice, moving down to join the mass in the fjord; and its rocks, too, are striated in the length of the valley, or from south to north. The moraine of Fondalen is a proof that a glacier once descended to the Holands Fjord at that point. Further evidence is found in the fact, that the sides of the valley are ground and striated for 700 feet and more above its bottom. Moreover, these polished and scored rocks can be traced up to and underneath the glacier. I crept for some yards under the ice, and found the floor of gneiss on which it rested smoothly polished and covered with scorings of all sizes, exactly the same in every respect as those high on the sides of the valley, in the fjord below, and away on the outer islands and skerries. Over this polished surface trickled the water of the melted ice, washing out sand and small stones from under the glacier.

We climbed the steep eastern side of the valley above the foot of the glacier, and found the hummocks of gneiss wonderfully glaciated up to a height of fully 700 feet. The gnarled crystalline rock has been ground away smoothly and sharply, so as to show its twisted foliation as well as the patterns of a marble are displayed on a polished chimney-piece. Even vertical or overhanging faces of rock are equally smoothed and striated. Many of the *roches moutonnées* are loaded with perched blocks of all sizes, up to

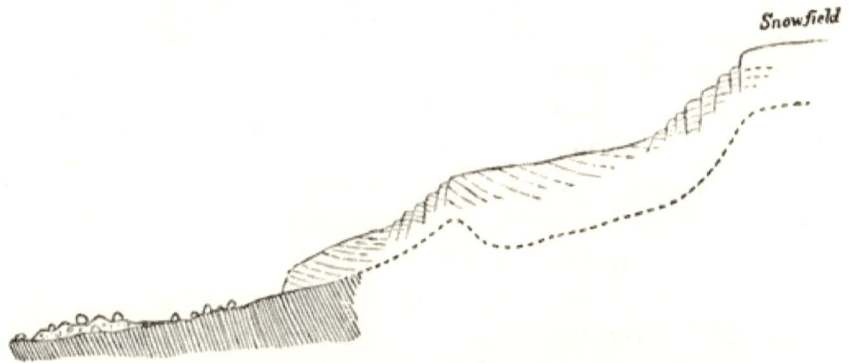

Fig. 10.—Longitudinal Section of smaller Glacier. Fondalen.

masses 30 or 40 feet long. Above the limit to which we traced the work of the ice the rocks begin to wear a more rugged surface, until along the summit of the ridges they rise into serrated crests and pinnacles. This rougher outline is of course the result of atmospheric waste, guided by the geological structure and chemical composition of the rocks.

The glacier descends from the snow-field, which we guessed to have there an elevation of about 3500 feet, to a point in the valley about 400 feet above the sea. The distance from the snow-field to the foot of the glacier looks not much more than one English mile—at least it is but short compared with the rapidity of descent. Hence the glacier is steep, and in some places much crevassed. Issuing from the upper snow in a steep, broken, and jagged slope of blue ice, it descends by a series of steps, till, getting compacted again in the valley below, it passes into a solid, firm glacier, with a tolerably smooth surface, forming a declivity of $12°$ or $15°$ (Fig. 10). At a point about half a mile or less from the foot of the glacier the valley suddenly contracts, and the glacier, much narrowed and compressed, tumbles over a second steep declivity in a mass of broken ice. The crevasses speedily unite, and after another descent of 300 or 400 yards at an angle of $25°$, the glacier comes to an end. At the point where the strangulation takes place the glacier lies in a kind of basin, of which the lower lip presents proofs of the most intense erosion. On the western bank, in particular, a mass of the mountain side which projects into the ice has been ground away, and shows plainly enough, by its form and striæ, that the glacier, ascending from the basin, has climbed up and over this barrier, so as to tumble down its northern or seaward side.

The course of this little glacier is now too short to admit of the formation of moraines. Yet there are large heaps of rubbish and enormous masses of rock scattered over the valley below; and the moraine at Fondalen is a further proof that, when the ice formerly filled the valley, its surface received abundant detritus from the mountain slopes on either side.

Opposite Fondalen, the Holands Fjord, passing through a deep and narrow channel on its northern bank, trends in an east-north-easterly direction; but just before taking this course it sends eastward a bay which terminates at the

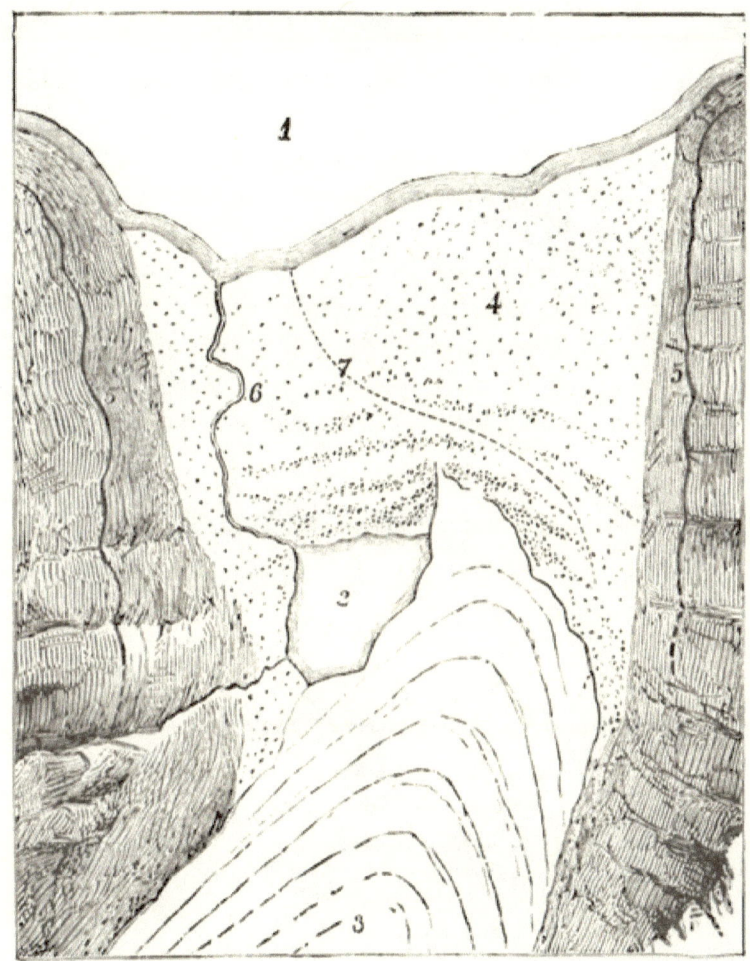

Fig. 11.—Sketch-map of lower end of larger Glacier. Fondalen.

1. Holands Fjord. 2. Small lake at end of glacier. 3. Glacier. 4. Alluvial plain rising into moraine mounds as it approaches the foot of the glacier. 5. Line of marine terrace or "raised beach," about 250 feet above sea-level. 6. Present course of stream. 7. Old course of stream.

mouth of a valley about a mile above the hamlet. This valley is considerably larger than that just described, and it is occupied by a much longer and larger glacier. To one who looks up the valley from the opposite side of the fjord, it seems as if the ample glacier which fills up the bottom sweeps down from the snow-field in a rapid descent to the very edge of the sea (Fig. 9). On a visit to the locality, however, it is found that between the foot of the glacier and the sea-margin there lies a plain of shingle and alluvium, partly covered with a brushwood of birch, and partly

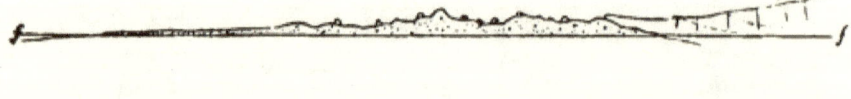

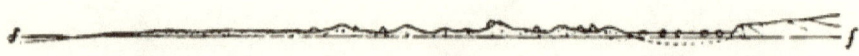

Fig. 12.—Sections across the lower end of the larger Glacier. Fondalen.

In the upper section, the glacier is shown overriding its moraine; in the lower, the small lake with floating ice intervenes between the end of the glacier and the moraine. In each section ƒƒ marks the level of the fjord.

with a scanty pasturage (Fig. 11). Near the ice the ground rises into ridges and hummocks, which increase in size towards the glacier. These are true moraine mounds, rising often 60 or 70 feet above their base, consisting of earth and stones, and strewn with large blocks of gneiss, porphyry, limestone, and other crystalline rocks. About a quarter of a mile from the margin of the fjord, along the eastern half of the breadth of the valley, these mounds come in contact with the foot of the glacier, which is there pushed in a long tongue down the valley. The ice overrides the moraine heaps, ploughing them and pushing them over (Fig. 12). On the west side of this prolongation of the glacier the ice is separated from the moraine mound by a small lake, of which the surplus waters find their way seaward by cutting through the moraine. Like many lakes still existing in Britain, this sheet of water is formed by the dam of rubbish thrown down by the glacier across the valley. It is full of fragments of ice, which break off from the parent mass, and float across to the north or lower side, where they strand on the moraine heaps, and gradually melt away. The smaller pieces, however, often find their way into the stream by which the lake discharges itself, and are then carried down into the fjord. From the mean of several observations taken with the aneroid, I estimated the surface of this lake to be about 25 feet above the level of high water in the fjord. We had no means of measuring its depth, yet, from the slope of the glacier, it may be inferred that the bottom of the ice is probably lower than the level of the sea.

Proofs that the glacier was once much larger than it is now may be well seen on the west side of the valley, a little above the lake. The shelving slopes of the mountain for several hundred feet upward have been shorn smooth, grooved, and striated, and every polished hummock of rock is loaded with huge fragments of stone and heaps of earth and angular rubbish. Here, as at every glacier we visited, the glaciation of the rocks, down to the minutest detail, was exactly similar to that of the coast and outer islets, as well as to that of the Scottish glens and sea-lochs.

But the feature which most interested us was the relation of this large glacier of Fondalen to the marine deposits of the locality. The foregoing sketch-map (Fig. 11) shows that the high terrace so marked along the sides of the

Holands Fjord enters this valley, and extends on the mountain sides, as far as, at least, the foot of the glacier. Hence the gravelly plain and the moraine mounds that separate the glacier from the fjord are overlooked on either side by a raised sea-beach. In examining attentively the nature of the material of which the mounds nearest the glacier were composed, we were struck with the difference between it and the loose, coarse character of the ordinary moraine rubbish, and its resemblance to the upper boulder-clay of Scotland. The glacier is pushing great noses of ice into and over these mounds, so that freshly-exposed sections are abundant. The deposit is a loose sandy clay or earth full of stones, among which the percentage of striated specimens is not large. The larger blocks of gneiss and schist appeared to us not to occur in this clay, but to be tumbled down upon it from the surface of the glacier. We had hardly begun to look over a surface of the clay ere we found fragments of shells, and in the course of a few minutes we picked up several handfuls, chiefly of broken pieces of *Cyprina Islandica*, but including also single valves of *Astarte compressa*, etc. We even took out two or three fragments which were sticking in the ice of the glacier. These shells were not peculiar to one spot, but occurred more or less abundantly across the valley.

From the nature of the material of which these mounds consist, and from the occurrence of marine shells, it was evident that we were looking not merely upon ordinary moraine heaps—the detritus carried down on the surface of the ice and discharged upon the bottom of the valley. The glacier was engaged in ploughing up the marine sediment which had been formerly deposited upon the submerged floor of the valley, and on the heaps of earth and clay now torn up were thrown the gravel and blocks brought down by the present glacier. In short, we saw here actually at work a process of excavation, by which it had been conjectured that the marine drift was removed from certain valleys in the British Isles.[1]

We made two attempts, both unsuccessful, to climb to the vast tableland of snow from which these glaciers are fed. But we succeeded in reaching a point from which a good view of the seemingly boundless undulating plain of smooth snow could be obtained. We ascended the ridge that separates the two glacier valleys just described. After leaving the raised beach of Fondalen, with its massive erratics, we climbed a steep slope, clothed with a thick brushwood of birch, mountain-ash, and dwarf-willow, and luxuriant masses of ferns, bilberries, cloudberries, juniper, rock-geranium, lychnis, etc. The beech trees are often a foot or a foot and a half in diameter at the base, and are the building material used at the hamlet of Fondalen below. These trees, at the height of 1320 above the sea, still often measure a foot across near the root, and 15 or 20 feet in height. At this height, and even considerably lower, there were large sheets of snow on the 12th of July, and these increased in number and depth as we ascended. The birch trees grow smaller and more stunted as they struggle up the bare mountain ridge, until they become mere bushes. The willows, in like manner, dwindle down till they look like straggling tufts of heather, though still bearing their full-formed catkins. At a height of 1690 feet, these stunted bushes at last give place to a scrub of bilberry, mosses, and lycopods. The mountain consists of gneiss, sometimes massive and jointed, sometimes fissile and flaggy, with a strike towards W., 15° S. The extent to which the higher

[1] See Sir A. C. Ramsay, *Glaciers of Switzerland and Wales*, 2d edition, p. 60.

rocky scarps have suffered from the disintegrating effects of the weather arrests attention; for the gneiss is split up along its joints into large blocks, which lie piled upon each other in heaps of angular ruin. We noticed one or two masses, differing in lithological character from the rocks around, and possibly ice-borne from some of the neighbouring eminences. On reaching a point 2700 feet above the fjord, our farther passage was arrested by a narrow, shattered, knife-edge of gneiss, along which, without suitable climbing gear, it was impossible to advance. But from this elevated point we could judge of the general aspect of the great snowy tableland of the Svartisen, which was sloping towards us, while the two glaciers were spread out in plan beneath.

The branch of the Holands Fjord which, opposite to the hamlet of Fondalen, strikes off to the north-east for seven or eight miles, is bordered on the south side, and closed in at its farther end, by a range of steep, almost precipitous, walls of rock, the summits of which are on a level with, and indeed form part of the great tableland. Here, as in so many other parts of Norway, we are reminded that the fjords are, after all, mere long sinuous trenches, dug deeply out of the edge of a series of elevated plateaux. And, looking up to the crest of these dark precipices, we see the edge of the high snow-plain peering over, and sending a stream of blue glacier ice down every available hollow. We counted seven of these tiny glaciers, exuding from under the snow, and creeping downward under the sombre cliffs of gneiss. Not one of them comes much below the snow-line, and none, of course, reaches the sea. The largest of them is near the end of the fjord, and appears as a broken, crevassed mass of ice, moulded as it were over the steep hillside, and, when seen from below, seeming about to slip off and plunge into the fjord. Fragments of it are continually breaking away, and rolling, with the noise of thunder and clouds of icy dust, down the shelving sides of the mountains. These glaciers are, for the most part, continuous with the snow-field, of which they are the icy drainage. One or two, however, lie in corries, quite detached from the main snow-field, though connected with it by continuous snow in winter.

The bright sunny Arctic nights led us not unfrequently and almost unconsciously to prolong the work of one day into the next. Once, at midnight, while sketching at Fondalen, I was amused by the loud and persistent call of a cuckoo perched on one of the neighbouring trees. The native non-migratory birds are evidently used to the ways of the sun in the Arctic summer, and, like the human population, know when to go to rest. But the tourist cuckoo was evidently quite unaware of the lateness of the hour, and continued his "twofold shout" as lustily as if it had been midday.

We left this delightful fjord not without regret, and catching again the coasting steamer at Melövaer, proceeded northwards. Between Melövaer and Bodö, the higher mountains have wonderfully craggy and spiry outlines, only their lower parts showing the smoothed contour of glaciation. But where the coast hills sink, as towards a fjord or bay, the ice-moulded forms can be traced to a greater height. To the north of Bodö, the contrast between the sharp weather-worn peaks above and the flowing ice-worn hummocks and hillsides below is singularly startling. Principal Forbes, who gave a characteristically faithful drawing to illustrate this feature, places the upper limit of glaciation at from 1500 to 2000 feet.[1] We should have

[1] *Norway and its Glaciers*, p. 58.

estimated it to be considerably lower. Through narrow kyles and intricate sounds, reminding one at every turn of detached portions of West Highland or Hebridean scenery, the steamer slowly wound its way, and then across the Vest Fjord to the Lofodden Islands. The weather now unfortunately proved unfavourable for geological observation. In sailing through the Rafte Sund we saw what looked like moraines at the mouths of some of the valleys, and the lines of moraine terraces continued as marked as ever. Rocks well ice-worn were also observed at the openings of some of the valleys, but we were rather impressed with the general ruggedness and absence of glaciation among the Lofoddens.

To the north of Tromsö lies the island of Ringvatsö, noticed by Mr. R. Chambers.[1] The moraine which he describes as damming up the circular sheet of water, whence the island takes its name, really coincides with the line of the higher of the two strongly-marked terraces or sea-margins of this part of the Norwegian coast. It thus illustrates the history of the moraine and terrace, below the smaller glacier at Fondalen. It was further interesting to mark that the glacier at Ringvatsö, partially hidden under snow, lies in a hollow or corry surrounded with precipices, and quite cut off from any snow-field. The accumulation of snow in the corry itself must thus be sufficient to give rise to the glacier. In looking at this island, I was forcibly reminded of the history of the glaciers of Tweedsmuir and Loch Skene in Peeblesshire and other old glacier grounds in Scotland, where, on dimples of the hill-tops, and in deep cliff-encircled recesses, snow enough gathered to form streams of ice, which caught and carried on their surface piles of rubbish and huge blocks of rock. A large snow-field is not neces-

[1] *Tracings of the North of Europe*, p. 145.

sary for the production of a glacier that may form comparatively extensive moraines.[1]

The south-western side of the Lyngen Fjord is formed by a mass of high ground, which shoots up steeply from the sea to a height of 4000 feet or more. Every hollow and cliff is smothered with snow, which descends in straggling streaks and patches almost to the edge of the water. We sailed up the fjord for some miles, and had a full view of this truly magnificent coast-line. We counted from ten to twelve small glaciers nestling in separate corries, and also two or three on the north-eastern side. There was here the same evidence of the formation of glaciers in small independent hollows of the mountains, quite detached, at least in the summer, from any large snow-field.

We halted at the island of Skjaervö (lat. 70°) for the purpose of making an excursion across the Kvenangen Fjord and up the Jökuls Fjord, to see the glacier which was said to reach the level of the sea[2] (Fig. 13). The metamorphic rocks among which the Jökuls Fjord lies are

---

[1] North Wales presents a number of illustrations of this remark, such as Cwm Graianog, Cwm Idwal, etc. (see Sir A. C. Ramsay's *Glaciers of North Wales*).

[2] This glacier was noticed by Von Buch, and is mentioned by Principal Forbes. When we visited it, I was not aware that a brief account of it had been given in vol. ii. of *Peaks, Passes, and Glaciers*, second series. Mr. J. F. Hardy, the writer of that description, started overland from Talvik on the Alten Fjord, and reached the Jökuls Fjord below the glacier, to which he ascended by boat. Like my own party, he did not climb the glacier, but he seems to have regarded it as connected with the snow-field above. Though I did not succeed in ascending the rugged cliffs, I had no doubt that the lower glacier, from its colour and the steepness and contraction of the gorge above it, is a true *glacier remanié*, and like the Suphelle glacier described by Forbes (*Norway and its Glaciers*, p. 149), is quite disconnected, at least in summer, from the snow-fields above.

for the most part of a flaggy quartzose character. Sometimes, especially where most fissile, they are violently crumpled. Parts of them pass into hornblende rock and actinolite schist. Their average strike is on an east and

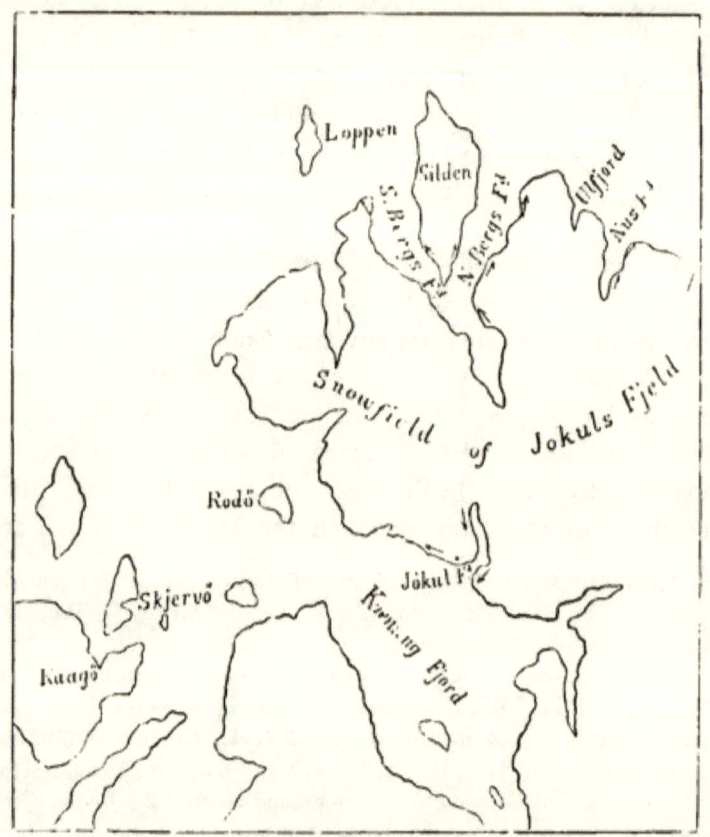

Fig. 13.—Map of the Jökuls Fjeld promontory (after Munch). The arrows show the direction of the old ice striæ.

west line. They are much jointed, and yield freely to the action of the weather. Hence, a rough and angular surface has very generally replaced the ice-moulded outlines, though these instill here and there remain. Numerous ancient mare terraces, especially the same two prominent ones already mentioned, may be traced along the sides of the Jökuls Fjord. The lower of these runs at a level of about 60 feet, the higher at about 152 feet (aneroid measurement) above high-water mark. The upper is especially marked, often running as a shelf cut out of the rock. This feature was noticed along many parts of the Norwegian coast, even (as in the Jökuls Fjord) in sheltered places where wave action cannot be supposed ever to have been very strong. As the date of these rock-terraces probably goes back into the glacial period, it occurred to me that they may have been due in large measure to the effects of the freezings and thawings along the old "ice-foot," and to the rasping and grating of coast ice. Such, too, may have been the origin of the higher horizontal rock-terraces of Scotland.

At the head of the fjord the terraces disappear along the steep bare sides of the mountains. A moraine mound of loose rubbish and large blocks lies on the west side, and extends a little way into the fjord, pointing towards a similar ridge on the opposite side, as if both were parts of a curved terminal moraine. The view from this ridge is singularly imposing. The sombre, precipitous mountains sweep upward from the edge of the water, seamed everywhere with streaks and sheets of snow. Down even to the beach these snow-drifts lie; and it gives a vivid impression of the high 'atitude of the place, that even in July there should be deep masses of snow overhanging tangle-covered rocks, and undermined by the wash of the waves. Over the crest of the mountains, at the head of the fjord, we see the edge of the great snow-field of the Jökuls Fjeld, and stealing down from underneath the snow comes a broken, shattered mass of glacier ice, broadest at the top, and narrowing downwards till its point disappears in a deep cleft or ravine, perhaps a

Fig. 14.—View of Jökuls Fjord Glacier.

third of the way down from the surface of the snow-field to the sea. The eastern part of this glacier seems plastered, as it were, over the forehead of the mountain, and is ever sending off fragments down the dark precipice below. Indeed, the whole glacier is in constant commotion, cracking and crashing and discharging masses of ice and snow, which pour over the black rocks in sheets of white dust, with a noise like the unintermitted thunder of a battle. These ice-falls are in large measure intercepted at the point where the glacier disappears behind the side of the ravine. They seemed, indeed, to collect in the ravine, and to slide down through it; for at its lower end a second glacier begins, and expands with the expansion of the hollow in which it lies, till it reaches the edge of the fjord, where it may be a quarter of a mile broad. This lower glacier appeared to me not connected with the snow-field, but a true *glacier remanié*, deriving its materials entirely from the avalanches of snow and ice that pour down upon its surface from the precipices overhead. It has a white, or dull greenish white colour, varied with well-marked dirt-bands. The slope of its surface was judged to be fully 20° or 25°. A few longitudinal crevasses make their appearance along the middle of the glacier, and a little farther down the transverse crevasses increase in number and size, until at its foot the glacier, broken by large semicircular rents, becomes a tumbled mass of ruin. These cliffs of granular loose-textured ice were observed in some places to overhang the waves. But the dark rock was likewise seen peering out along the water's edge, underneath the ice, which does not push its way out to sea in a mass, but ends abruptly where it meets the water. From these icy walls small fragments and large slices break off, and fall either on the margin of rock or into the fjord, which is thus covered with hundreds of miniature icebergs, slowly drifted downwards against wind and tide, by the surface current of fresh water (Fig. 15). This process is called "calving" by the natives, and so great is the commotion sometimes produced that, according to the information collected by Von Buch, the Lapp huts along the margin of the fjord are sometimes inundated by the waves propagated outwards from the falling masses. The floating fragments of ice look like little models of Arctic bergs, with forms often singularly fantastic. They may be seen shifting their position, and even capsizing, as their submerged parts melt away; some of them carry stones and earth on their surface, while many, aground

Fig. 15.—Section of Foot of Jokuls Fjord Glacier.

along the margin of the fjord, rise and fall with the tide or with the ripple of the waves. We passed two or three which were from 8 to 10 feet long, and rose from 3 to 4 feet out of the fjord. Our boat grated against several which seemed only a foot or two in size, yet the shock of the collision showed how much larger was the portion concealed under water.

To the east of the upper glacier the snow-field sends another icy stream down the face of the shelving precipices which descend into a higher valley. We could hear the roar of the avalanches even when the glacier itself was hidden behind the intervening mountain-spur. From the rocky declivities of the Jökuls Fjord also stones were heard and seen bounding from point to point in their descent towards the long heaps of *débris* at the bottom. In short, in this lonely uninhabited spot, the activity and ceaselessness of the wasting powers of nature come before the traveller with a memorable impressiveness. The wide snow-field that seems to lie so sluggish and still among the distant mists, is yet seen to be in slow but constant motion, pushing its ice-streams towards the valleys, and grinding down the hard rocks over which it moves. Frosts, rain, and springs have scarped the shoulders of every mountain, and poured long trains of rubbish down its sides. And if this can be now done under the present climate of Norway, how much more powerful must the abrasion have been when the ice, in place of being arrested on the brow of the mountain, filled up the fjord, and pushed its way into the Arctic Sea!

From the open mouth of the Kvenangs Fjord, in the passage between Skjaervö and the Jökul, the outline of the neighbouring land is well seen. The steep, serrated ridge of the Kvenangs Tinderne shows its tiny glaciers nestling in corries both on its northern and southern slopes. The sides of the Kvenangs Fjord are ice-moulded and striated in the direction of the inlet, and its islands are only large *roches moutonnées*. In looking back at the mountainous track of the Jökuls Fjeld, we see that it is another snowy tableland jutting out as a promontory into the Arctic Sea, deeply trenched with long, narrow fjords, and pushing glaciers down every glen and hollow that descends from the plateau of snow. I sketched these scenes at midnight, when the sun, after gathering round him the crimson and orange glories of his setting, lingers along the northern horizon, and then spreads over the sky the tender hues of sunrise—a blending of sunset and dawn which is one of the most memorable experiences of travel in the north.

We visited the north-western and northern sides of this snow-field, boating up the Bergs Fjord, to the hamlet of that name, and after ascending to its glaciers, continuing our excursion by boat into the Nus Fjord. (See Map, Fig. 13.) In ascending the South Bergs Fjord, we found the gneissic and schistose rocks polished and striated from east to west, which is the direction of the inlet, and in turning into the North Bergs Fjord, which runs nearly at a right angle to the other, the striæ were seen to turn out of the Lang Fjord and bend northward through the northern limb of the Bergs Fjord. At the hamlet of Bergsfjord these ice-mouldings are especially well shown, and there, as well as along many parts of the fjord, occur lines of rock-terrace, often strewed with quantities of angular blocks. Two of the most marked of these horizontal bars have an elevation of about 50 and 150 feet respectively. Behind the hamlet the ground slopes up to a point about 250 feet above the sea, beyond which lies the mouth of a valley that runs up into the heart of the mountains. We climbed the terraced slope leading to this recess, and found that the lower half of the valley is occupied by a lake about a mile long, and said to be 30 fathoms deep. It lies in a rock basin, and the rocks around its margin show that they have been powerfully abraided by ice. We were told that three weeks before our visit this lake was solidly frozen over; great sheets of snow, indeed, still descended to the water's edge, and were melting away under the glare of a fierce July sun. At the far end of the valley mounds of angular rubbish, cumbered with huge blocks of stone, stretched from side to side, while overhead two glaciers came out of the edge of the snow-field, and hung down the steep mountain side

—the longer one almost reaching the bottom of the valley. We started a small herd of reindeer pasturing among the moraine heaps. The animals bounded over the snow-wreaths, always choosing the firmest portions which stretched as natural bridges across the stream that worked its way underneath. Here, too, the ice was ever breaking up, and crashing down the precipices in clouds of snowy dust. The *débris* of ice gathered into talus heaps below, like the *cones de dejection* at the foot of a winter torrent.

From Bergsfjord we continued our boating voyage down the fjord, and found fresh proofs that a vast body of ice, descending from the lofty Jökuls Fjeld, had moved northwards along the length of the inlet. Every promontory was beautifully smoothed and polished; while the grooves and striæ slanted up and over the projecting bosses of rock, as they do in Loch Fyne and the other western sea-lochs of Scotland. Round the headland at the mouth of the Bergs Fjord we turned eastward, and soon passed the mouth of the Ulfjord. We could see that, at the far end of that inlet, the snow of the great tableland moves outward to the edge of the dark precipices which encircle the Ulfjord, and actually forms on the crest of these precipices a white cliff, from which, of course, avalanches are constantly falling. Yet the under part of this snowy cliff is not snow, but ice, as shown by its blue colour contrasting with the whiteness of the upper layer, which is snow or *névé*. At the foot of the precipice a glacier, derived probably in part, like that of Jökuls Fjord, from the ice-falls from above, creeps towards, but does not reach, the bottom of the valley. Continuing our eastward journey, we saw the same terraces, still skirting the hillsides, now as green platforms of detritus loaded with angular blocks, and now as sharp horizontal notches in the bare rocks. We were likewise struck here, as in other parts of the Norwegian coast, with the greater freshness of the ice-markings near the sea-level, when compared with those higher up—a difference which is likewise very noticeable in the West of Scotland.

An incident occurred in this part of the journey which helped to strengthen the parallelism I had been tracing between the old glacial conditions of Scotland and those now existing in Arctic Norway. In one of the hospitable and solitary merchants' houses I found a little girl playing with valves of the red Iceland scallop (*Pecten Islandicus*) or "*röde heste*," red horses, as she called them. They were evidently recent, and not fossil shells, and I found them strewn plentifully on the beach. This species once lived abundantly among the western fjords of Scotland, and its valves are there plentiful in the upraised sea-floor of the glacial period. But it still flourishes in the fjords of Norway.

The Nus Fjord is about six miles long, and lies between the Ulfjord and Oxfjord. Its margin is terraced by the same horizontal lines so constant in this region. Its southwestern side presents a singularly Arctic scene. A range of deeply cleft and embayed crags and precipices, plentifully streaked with snow, rises up to the edge of the snowfield, which, as usual, sends down into every larger valley a stream of blue ice. Eight or ten distinct glaciers may be counted, of which at least three descend from the snowfield. The others lie in corries detached from the snowfield, though in some cases connected with it by nearly perpendicular streaks of snow. Here, as in the Ulfjord, the edge of the great sheet of snow which covers the tableland may be seen ending off abruptly as a cliff upon the crest of a dark precipice of rock, and from the colour of the lower part of the cliff it is plain that, from pressure and

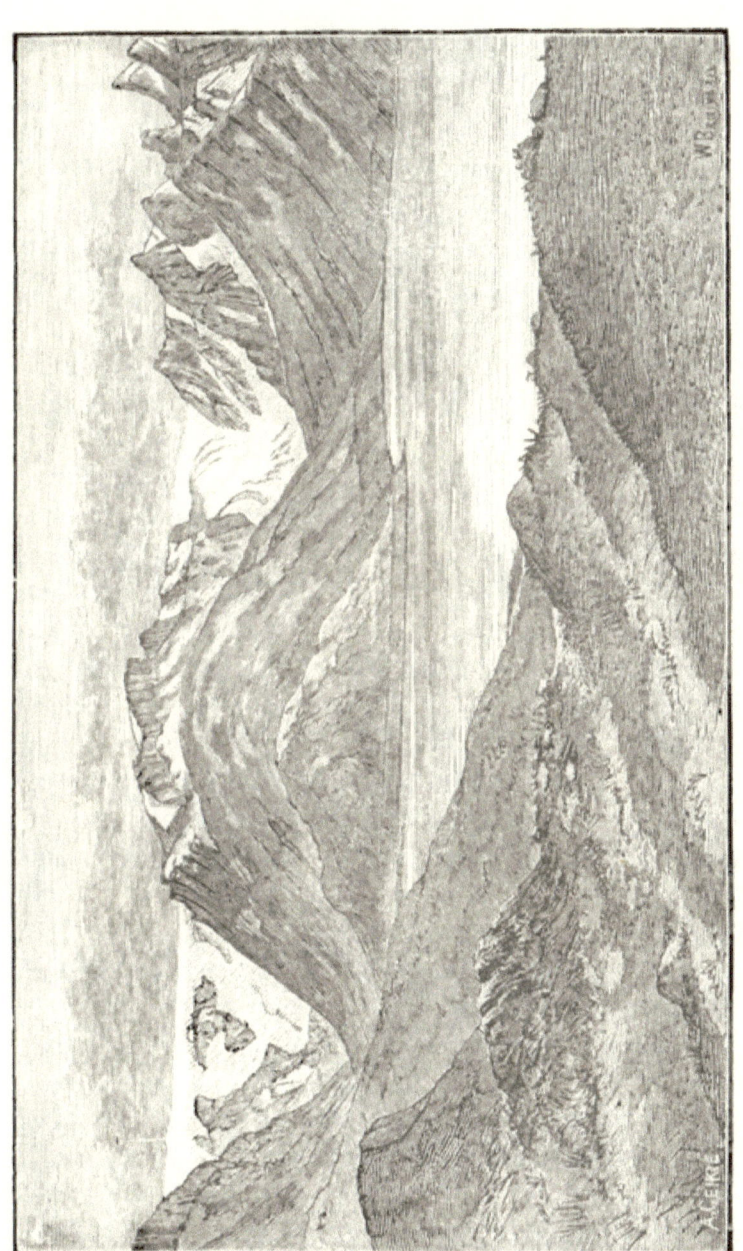

Fig. 16.—View of Glaciers at the head of Nus Fjord.

motion, the under portion of the snow-sheet is converted into ice, and as ice reaches the verge of the tableland, where it breaks sharply off, and sends its ruins to the bottom of the precipice underneath. There the *débris*, mingled with the winter snow, is anew converted into solid ice, and creeps downward as a glacier.

At the head of the fjord, on the south-east side, the mouth of a valley which terminates inland at the foot of a glacier is blocked up by an old moraine. Behind this rampart of detritus the valley spreads out as an alluvial plain, evidently at one time a lake formed by the moraine barrier at the foot. The moraine itself is strewed with enormous angular blocks of rock, beside which the huts of a miserable Lapp encampment look like mere pebbles. The side of this moraine facing the fjord is cut by the 50 foot beach. On the opposite side of the fjord a valley, at the head of which a glacier comes down from the Snee-fond, opens upon the shore, and is curtained across by a terrace, the surface of which is mottled with a number of irregular concentric mounds. We had no opportunity of examining these, but they seemed to be moraine heaps left by the glacier when it came down to the fjord. They vividly recalled the singular concentric mounds that overlie the terrace at the mouth of the old glacier valley of the Brora in Sutherlandshire.

We walked along the north-east side of the fjord, and found the rocky declivity terraced with old sea-margins, which run along like ancient and ruined roadways. They occur up to perhaps 200 or 250 feet above the sea-level, and are cut in the hard rock. They are covered with loose blocks, partly derived from the rocks around, but probably in part also transported from a higher part of the valley. On the beach we met with well ice-worn bosses of gneiss, slipping beneath a gray sandy clay full of Arctic shells—a conjunction which is closely paralleled by one on the shores of Loch Fyne (Figs. 17, 18). Both in the Norwegian and Scottish examples the rocks underneath are beautifully

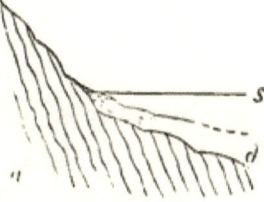

Fig. 17.—Section on beach at Nus Fjord.

d Sandy-gray clay, with *Tellina proxima, Saxicava rugosa, Astarte elliptica, Cyprina Islandica*, etc. *a* Ice-worn gneissose rocks. *s* High-water mark.

smoothed and grooved, showing that in each case the ice which moulded them moved down the length of the inlet. To the north and east of the Jökuls Fjeld the ground becomes lower, and descends wholly below the snow-line.

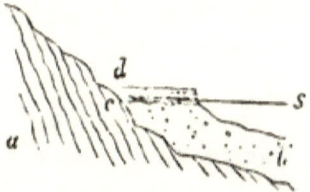

Fig. 18.—Section on beach at Ardmarnock, Loch Fyne.

d Sandy-gray clay, full of *Tellina proxima, Astarte borealis, Natica clausa, Cyprina Islandica* (in fragments sometimes seven-twelfths of an inch thick) and other northern shells. *c* Finely stratified red clay, without shells. *b* Boulder-clay. *a* Ice-worn gneissose rocks. *s* High-water mark.

The hills that bound the Alten Fjord, instead of rising into serrated peaks, like the higher tracts to the south, have a well ice-worn aspect, and recall the hills of Cantyre, or the scenery of parts of the Hebrides. Indeed, the whole of this northern district of Norway, from the Alten Fjord to beyond the North Cape, has the smoothed outline which farther southward is found only on the lower zone of the mountains.[1] It seems as if a sheet of ice, descending from the south, had overridden all the fjords here and the comparatively low hills between them, and had advanced northward into the Arctic Sea.

In fine, this short excursion into the northern part of Scandinavia furnished us with abundant proofs that the glaciation of the west of Norway was produced by a mass of land-ice, of which the present glaciers are the representatives. It likewise confirmed, in a most impressive way, the conclusion which has gained ground so rapidly within the last few years, that the glaciation of the Scottish Highlands, as well as of the rest of the British Isles, is in the main the work, not of floating bergs, but of land-ice. This conclusion may, indeed, be regarded as demonstrated beyond all cavil by the ice-marks of Norway. Much good work might be done by trying to work out a detailed comparison of the glaciation of the Scandinavian peninsula with that of this country. More especially would it be of importance to ascertain how far the glacial deposits of the two countries can be compared. Doubtless the drift-covered slopes of Sweden, and those of the east and centre of Scotland, must have many geological features in common. It will perhaps be found that some of the difficulties which our Scottish drift presents are explained by the more extensive deposits of the north, while the latter may likewise suggest new explanations of phenomena supposed to be already sufficiently intelligible.[2]

---

[1] We did not go farther than Hammerfest, but the same contour is retained over the low, tame district that separates Hammerfest from the North Cape.

[2] Since the publication of this paper in January 1866, much labour has been bestowed upon the glacial phenomena of Scandinavia and of Scotland.

VII.

A FRAGMENT OF PRIMEVAL EUROPE.[1]

WHEN the history of the growth of the European area is traced backward through successive geological periods, it brings before us a remarkable persistence of land towards the north. The stratified formations bear a generally concurrent testimony to the existence of a northern source whence much of their sediment was derived, even from very early geological times. In their piles of consolidated gravel, sand, and mud, their unconformabilities and their buried coast-lines, they tell of some boreal land which, continually suffering denudation, but doubtless at intervals restored and augmented by upheaval, has gradually extended over the whole of the present European area. The chronicles of this most interesting history are at best imperfect, and have hitherto been only partially deciphered. They naturally assume an increasingly fragmentary and obscure character in proportion to their antiquity. Nevertheless traces can still be detected of the shores against which the oldest known sedimentary accumulations were piled. These shores have of course been deeply buried under the deposits of subsequent ages. But the whirligig of time has once more brought them up to the light of day

[1] *Nature*, August 1880.

by stripping off the thick piles of rock beneath which they have lain preserved during so vast a cycle of geological revolutions. I shall here describe a fragment of this earliest land, and allude to some of the geological problems which it suggests.

In the north-west of Scotland, along the seaboard of the counties of Ross and Sutherland, a peculiar type of scenery presents itself, which reappears nowhere else on the mainland. Whether the traveller approaches the region from the sea or from the land, he can hardly fail to be struck by this peculiarity, even though he may have no specially geological eye for the discrimination of rock-structures. Seen from the westward or the Atlantic side, as, for example, when sailing into Loch Torridon, or passing the mouths of the western fjords of Sutherlandshire, the land rises out of the water in a succession of bare rounded domes of rock, crowding behind and above each other as far as the eye can reach. Not a tree or bush casts a shadow over these folds of barren rock. It might at first be supposed that even heather had been unable to find a foothold on them. Gray, rugged, and verdureless, they look as if they had but recently been thrust up from beneath the waves, and as if the kindly hand of nature had not yet had time to clothe them with her livery of green. Strange, however, as this scenery appears when viewed from a distance, it becomes even stranger when we enter into it, and more especially when we climb one of its more prominent heights and look down upon many square miles of its extent. The whole landscape is one of smoothed and rounded bosses and ridges of bare rock, which, uniting and then separating, inclose innumerable little tarns (Fig. 20). There are no definite lines of hill and valley; the country consists, in fact, of a seemingly inextricable labyrinth of hills and valleys, which, on the whole, do not rise much above, nor sink much below, a general average level. Over this expanse, with all its bareness and sterility, there is a singular absence of peaks or crags of any kind. The domes and ridges present everywhere a rounded, flowing outline, though here and there this outline has been partially defaced by the action of the weather.

The rocks that have assumed this external contour are the Archæan, Fundamental, Lewisian, or Laurentian gneiss, which, as Murchison showed, form the platform whereon the rest of the stratified rocks of Britain lie. They do not, however, cover the whole surface of these north-western tracts. On the contrary, they form a broken fringe from Cape Wrath to the Island of Raasay, coming out boldly to the Atlantic in the northern half of its course, but throughout the southern portion retiring chiefly towards the heads of the bays and sea-lochs, and even extending inland to the head of Loch Maree. The reason of this want of continuity is to be found in the spread of later formations over the gneiss. At the base of these overlying deposits comes a mass of dark red sandstone and conglomerate (classed as Cambrian by Murchison and his associates), which, in gently-inclined or horizontal strata, sweeps across the platform of gneiss, rising here and there into solitary cones or groups of cones fully 3400 feet above the sea. No contrast in Highland scenery is more abrupt and impressive than that between the ground occupied by the old gneiss and that covered by this overlying sandstone group. So sharp is the line of demarcation between the two tracts that it can be accurately followed by the eye even at a distance of several miles. Where the sandstone supervenes, the tumbled sea of bare gray gneiss is succeeded by smooth heathy slopes, through which the flat or gently-inclined parallel edges of the beds protrude in successive lines of terrace. As the ground rises into conical mountains, the covering of heather grows more and more scant, but the same terraced bars of rock continue to rise even to the summits, so that these vast solitary cones, standing apart on their platform of gneiss, have rather the aspect of rudely symmetrical pyramids than the free, bold sweep of crag and slope so characteristic of other Scottish mountains.

The depth of these sandstones must amount to several thousand feet. Even in single mountains a thickness of more than 3400 feet can be taken in at a glance of the eye from base to summit (Fig. 19). Yet when this massive formation is followed along the belt of country in which it lies it is found to thin out rapidly and even for some distance to disappear. Such a disappearance might arise either because the sandstone was not continuously deposited, or more probably because it was unequally worn down before the next group was accumulated upon it. Evidently the solution of this question has an important bearing on any reconstruction of the early geography of the region.

Above the red sandstones and creeping transgressively across them lies the deep pile of white quartzites, limestones, and schists, which Mr. C. W. Peach's discovery of recognisable fossils in them at Durness showed to be of Lower Silurian age. Another well-marked contrast of scenery is presented where these rocks abut upon those just described. The quartzites rise into long lines of bare white hills which, as the rock breaks up under the influence of the weather, are apt to be buried under their own *débris* even up to the summits. Here and there outlying patches of the white rock may be seen gleaming along the crests of the dark sandstone mountains, like fields of snow or nascent glaciers (Fig. 19). Quartzites, limestones, and schists dip

Fig. 19.—Ben Leagach, Glen Torridon, a mountain of red Cambrian sandstone, capped with white quartzite.

away to the east and pass under the vast series of younger schists which form most of the rest of the Scottish Highlands. This order of succession, first established by Murchison, can be demonstrated by innumerable lines of natural section. I have myself traced it through the mountainous country from Cape Wrath to Skye, and in many traverses across Sutherland and Ross. I have sought for evidence of the reappearance of the old or fundamental gneiss of the north-west, and have ransacked every Highland county in the search, but have never found the least trace of that rock beyond its limits in Sutherland and Ross. Its distinctive gneisses and other crystalline masses, so wonderfully unlike anything else in the Highlands, never reappear to the east. And that strange mammillated, bossy surface is found in the north-west alone.

To realise what the appearance of the old gneiss at the present surface involves we must bear in mind that it was first buried under several thousand feet of red sandstone, that the area was then further submerged until the vast pile of sediment was deposited out of which the Highlands have been formed, that these sedimentary accumulations—how many thousand feet thick we cannot yet tell—were subsequently over the Highland area crumpled and metamorphosed into crystalline schists, and that finally towards the west the ancient platform of gneiss was once more ridged up and gradually bared of its superincumbent load of rock, until now at length some portions of it have been once more laid open to the air.

There is thus a special historical interest in this fragment of the old gneiss country. It is a portion of the earliest European surface of which as yet we know anything—a surface in chronological comparison with which the Alps are of quite modern date. For many years past

Fig. 20.—View of the ancient platform of gneiss looking eastward from above Scourie, Sutherlandshire.

The hill to the left (one bird) is Arkle (2580 feet), composed of quartzite resting on the fundamental gneiss, and dipping eastward; the conical mountain Stack (2364 feet, marked here by two birds) consists of the old gneiss—the highest elevation reached by this rock on the mainland. The gneiss sweeps southward and underlies the great conical mountain of Quenaig (2653 feet, four birds), while in the distance (three birds) are seen the quartzite heights of Ben More Assynt (3235 feet).

I have at intervals wandered over it, finding in its undulations of bare rock a fascination which a fairer landscape might fail to exert. Each visit suggests some fresh problem, if it does not cast light on earlier difficulties. One of the questions which must particularly engage the attention of every observant traveller in Western Sutherland and Ross is the origin of that extraordinary contour presented by the gneiss. A very slight examination shows that every dome and boss of rock is ice-worn. The smoothed, polished, and striated surface left by the ice of the glacial period is everywhere to be recognised. Each hummock of gneiss is a more or less perfect *roche moutonnée*. Perched blocks are strewn over the ground by thousands. In short, there can hardly be anywhere else in Britain a more thoroughly typical piece of glaciation.

An obvious answer to the question of the origin of the peculiar configuration of this gneiss country is to refer it to the action of the last ice-sheet which covered Britain. That the gneiss was powerfully ground down by that ice is sufficiently manifest. But if the peculiar bossy surface is to be thus explained we are confronted by the difficulty that the ice must have acted far more effectively on the gneiss than on any other rock in the region. Yet there is nothing either in the structure of the rocks or in the configuration of the ground to make the erosion greater on the gneiss than on the red sandstone or quartzites and schists. The same side of a sea-loch may be seen to present slopes both of gneiss and sandstone; the gneiss is always worn into smooth domes, ridges, and hollows; but the sandstone retains its parallel bands of rocky terrace. The difference is evidently not due to any recent greater glacial abrasion of the gneiss. The area of high ground above the gneiss platform in Sutherlandshire is compara-

Fig. 21.—Ben Shieldag, Loch Torridon—a hill of flat Cambrian red sandstone resting upon an uneven hummocky surface of the old gneiss.

tively small. As shown in Fig. 20, it rises somewhat steeply from the west, its chief area and drainage lying towards the east. I have visited those tracts of the Highlands where the rocks approach nearest to the type of the ancient gneiss, and where the conditions have been most favourable for intense glaciation. No more promising locality for a comparison of this kind could be found than the deep defiles of Glen Shiel and Kintail. The rocks have there been extremely metamorphosed, and have been exposed to the action of ice descending from some of the highest uplands in the West of Scotland. Yet we look in vain among them for any semblance of the bare bossy surface of the old gneiss.

A further difficulty arises when we reflect that in the general erosion of the country the gneiss, being covered by later formations, would be the last to be attacked, and in so far as it was so covered, must have been exposed to the erosive action of the ice for a shorter time than the overlying rocks. We might therefore have presumed that instead of being more, it would have been less trenchantly worn down than these. Its great toughness and durability, which have enabled it to retain the ice-impress so faithfully, must have given it considerable powers of resistance to the grinding action of the glacier.

Every fresh excursion into these northern wilds has increased my difficulty in accounting for the peculiar contours of the gneiss ground by reference merely to the work of the Glacial Period. A recent visit, however, seems at last to have thrown some light on the matter. I had long been familiar with the fact that the platform of gneiss on which the red sandstones and conglomerates were laid down abounded in inequalities even at the time of the deposit of these strata. Its uneven surface rose here and there into

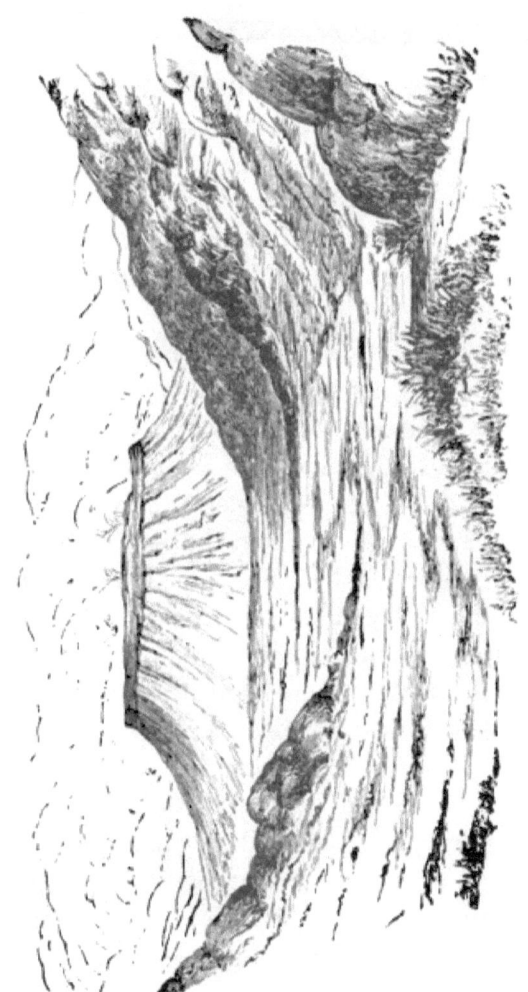

Fig. 22.—View of outlier of Cambrian breccia and sandstone among gneiss hills, near Gairloch.

high ridges and cones, of which Stack is a diminished representative, and sank into depressions now occupied by thick masses of sandstone. But I have lately observed that not only do these larger features pass under the sandstone, but that the minor domes and bosses of gneiss do so likewise. On both sides of Loch Torridon, for example, the hummocky outlines of the gneiss can be seen emerging from under the overlying sandstones (Fig. 21). On the side west of Loch Assynt similar junctions are visible. But some of the most impressive sections occur in the neigbourhood of Gairloch. Little more than a mile to the north of the church the road to Poolewe descends into a short valley surrounded with gneiss hills. From the top of the descent the eye is at once arrested by a flat-topped hill standing in the middle of the valley at its upper end, and suggesting some kind of fortification: so different from the surrounding hummocky declivities of gneiss is its level grassy top, flanked by wall-like cliffs rising upon a glacis-slope of *débris* and herbage (Fig. 22). Closer examination shows that the little eminence is capped with a coarse reddish breccia made up of fragments from the surrounding gneiss. The stones in this deposit are for the most part perfectly angular, and are sometimes stuck on end in the mass. They underwent but little re-arrangement after they were thrown down, though occasional lenticular seams of red sandstone running through the rock serve to prove that it is lying as a flat cake on the gneiss. My friend Mr. Norman Lockyer accompanied me in the examination of this hill. We searched long for a striated stone among the component materials of the breccia, but the matrix was too firm to allow us to bare and extract any of the pebbles or boulders. We traced, however, the characteristic rounded bossy surface of the gneiss until it passed under the breccia, and were

VII]   A FRAGMENT OF PRIMEVAL EUROPE.   157

convinced that, could the outlier of breccia be stripped off, the same kind of surface would be found below it as on the gneiss above and around. The valley containing this little fragment of a once more extensive deposit of breccia certainly existed as a hollow in Cambrian times. From the narrowness of its present outlet, which has been cut by the escaping streamlet, and from the nature of the breccia, we may infer with some plausibility that the hollow was filled with water, and may have been a lake. It was almost certainly a rock-basin, surrounded with hills of gneiss that had been worn into undulating dome-shaped hummocks.

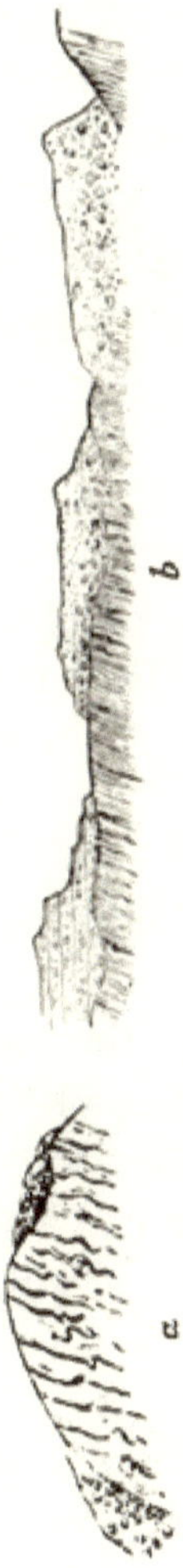

Behind the new hotel at Gairloch the ground rises steeply into a rocky bank of the old gneiss. Along the base of these slopes the gneiss (which is here a greenish schist) is wrapped round with a breccia of remarkable coarseness and toughness. We noticed some blocks in it fully five feet long. It is entirely made up of angular fragments of the schist underneath, to which it adheres with great tenacity. Here again rounded and smoothed domes of the older rock can be traced passing under the breccia, as at *a* in Fig. 23. On the coast immediately to the south of the new Free Church a series of instructive sections lays bare the worn undulating platform of gneiss, with its overlying cover of coarse angular breccia (*b*, Fig. 23). Similar evidence occurs to the north of Loch Inver.

Fig. 23.—Sections of the junction of the fundamental gneiss and overlying Cambrian breccia. Gairloch.

On these far northern shores, then, there still remain fragments of the surface on which our oldest sedimentary accumulations were deposited. These fragments are found to bear in their smooth hummocky contours a striking resemblance to the surface which geologists now always associate with the action of glacier-ice. There can at least be no doubt that they are denuded surfaces. The edges of the vertical and twisted beds of gneiss and schist have been smoothly bevelled off. These rocks, however, would never have assumed such a contour if exposed merely to ordinary sub-aërial disintegration. They would have taken sharp craggy outlines like those which are here and there gradually replacing the ice-worn curves of the *roches moutonnées*. They have certainly been ground by an agent that has produced results which, if they were found in a recent formation, would, without hesitation, be ascribed to land-ice. The breccia, too, is quite comparable to moraine stuff. Without wishing at present to prejudge a question on which I hope yet to obtain further evidence, I think we have in the meantime grounds for concluding that in the north-west of Scotland there is still traceable a fragment of the earliest known land-surface of Europe, that this primeval country had a smooth undulating aspect not unlike that of the west of Sutherland at the present time, that it contained rock-hollows, some of them filled with water, that into these hollows piles of coarse angular detritus were thrust, that around and beneath the tracts where this detritus accumulated the gneiss was worn into dome-shaped forms strongly suggestive of the operation of land-ice, and that though the ice of the last Glacial Period undoubtedly ground down the platform of gneiss, bared as it was of the overlying formations, it found a surface already worn into approximately the same forms as those which it presents to-day.

VIII.

ROCK-WEATHERING MEASURED BY THE DECAY OF TOMBSTONES.[1]

A BUILDING or other object having an antique aspect is called "age-worn" or "time-eaten," or is described by some other phrase which implies that during a long course of years the object in question had been suffering from some slow kind of change. We speak of "the gnawing tooth of time," as if time were a material form, or at least a force or energy endowed with certain powers of destructiveness, though obviously mere lapse of time can have no such influence. That there is some close connection, however, between antiquity and decay is manifest on every side. An ancient building is expected to look more or less decayed: if we find it to look fresh, we immediately, and as it were instinctively, doubt its antiquity.

The change which in course of time results in producing the crumbling, venerable aspect of a piece of old human architecture is but part of the continual change in progress upon natural surfaces of rock all over the world. The cliffs of a mountain-side or sea-shore reveal precisely the same alteration, but in a higher degree, for they rise on a more stupendous scale and have been exposed to the

[1] *Proceedings of the Royal Society of Edinburgh*, 1880.

weather for an enormously longer time than even the oldest of human erections.

This kind of decay is briefly described as "weathering." It is a complex process, however, or rather a series of processes, depending on the one hand upon the relative efficiency of changes of temperature, wind, rain, and frost; on the other hand, upon the composition and texture of the stone itself. Apart from the problem of the nature of the change lies the question of its rate. Actual time-measures are as yet so few in geological inquiry that any attempt may be welcomed which promises to supply one. The rate of weathering of rocks appears to be a question in which precise measurement should not be by any means unattainable. Comparatively little, however, has yet been done to determine with precision or even approximately, the rate at which the exposed surfaces of different kinds of rock decay. A few years ago, some experiments were instituted by Professor Pfaff of Erlangen to obtain more definite information on this subject.[1] He exposed to ordinary atmospheric influences carefully measured and weighed pieces of Solenhofen limestone, syenite, granite (both rough and polished), and bone. At the end of three years he found that the loss from the limestone was equivalent to the removal of a uniform layer 0·04 mm. in thickness from its general surface. The stone had become quite dull and earthy, while on parts of its surface fine cracks and incipient exfoliation had appeared. The time during which the observations were continued was, however, too brief to allow any general deductions to be drawn from them as to the real average rate of disintegration. Professor Pfaff relates that during the period a severe hailstorm broke one of the plates of stone. An exceptionally powerful cause of this

[1] *Allgemeine Geologie als exacte Wissenschaft*, p. 317.

nature might make the loss during a short interval considerably greater than the true average of a longer period.

It occurred to me recently that data of at least a provisional value might be obtained from an examination of tombstones freely exposed to the air in graveyards, in cases where their dates remained still legible or might be otherwise ascertained. I have accordingly paid attention to the older burial-grounds in Edinburgh, and have gathered together some facts which have, perhaps, sufficient interest and novelty to be worthy of publication.

At the outset it is of course obvious that in seeking for data bearing on the general question of rock-weathering, we must admit the kind and amount of such weathering visible in a town to be in some measure different from what is normal in nature. So far as the disintegration of rock-surfaces is effected by mineral acids, for example, there must be a good deal more of such chemical change where sulphuric acid is copiously evolved into the atmosphere from thousands of chimneys than in the pure air of country districts. In these respects we may regard the disintegration in towns as an exaggeration of the normal rate. Still, the difference between town and country may be less than might be supposed. Surfaces of stone are apt to get begrimed with dust and smoke, and the crust of organic and inorganic matter deposited upon them may in no small measure protect them from the greater chemical activity of the more acid town rain. In regard to daily or seasonal changes of temperature, on the other hand, which unquestionably exert a powerful influence in the disintegration of rock-surfaces, any difference between town and country may not impossibly be in favour of the town. Owing, probably, to the influence of smoke in retarding radiation, thermometers placed in open spaces in town commonly mark an extreme nocturnal temperature not quite so low as those similarly placed in the suburbs, while they show a maximum day temperature not quite so high.

The illustrations of rock-weathering presented by city graveyards are necessarily limited to the few kinds of rock employed for monumental purposes. Around Edinburgh the materials used are of three kinds:—1st, Calcareous, including marbles and limestones; 2d, sandstones and flagstones; 3d, granites.

I. CALCAREOUS.—With extremely rare exceptions, the calcareous tombstones in our graveyards are constructed of ordinary white saccharoid Italian marble. I have also observed a pink Italian shell-marble, and a finely fossiliferous limestone, containing fragments of shells, foraminifera, etc.

In a few cases the white marble has been employed by itself as a monolith in the shape of an obelisk, urn, or other device; but most commonly it occurs in slabs which have been tightly fixed in a framework of sandstone. These slabs, from less than one to fully two inches thick, are generally placed vertically; in one or two examples they have been inserted in large horizontal sandstone slabs or "through-stanes." The form into which the stone has been cut, and the position in which it has been erected, have had considerable influence on its weathering.

A specimen of the common white marble employed for monumental purposes was obtained from one of the marble-works of the city, and examined microscopically. It presented the well-known granular character of true saccharoid marble, consisting of rounded granules of clear transparent calcite, averaging about $\frac{1}{100}$ of an inch in diameter (Fig. 24, A). Each granule has its own system of twin lamellations, and interference colour-bands. The funda-

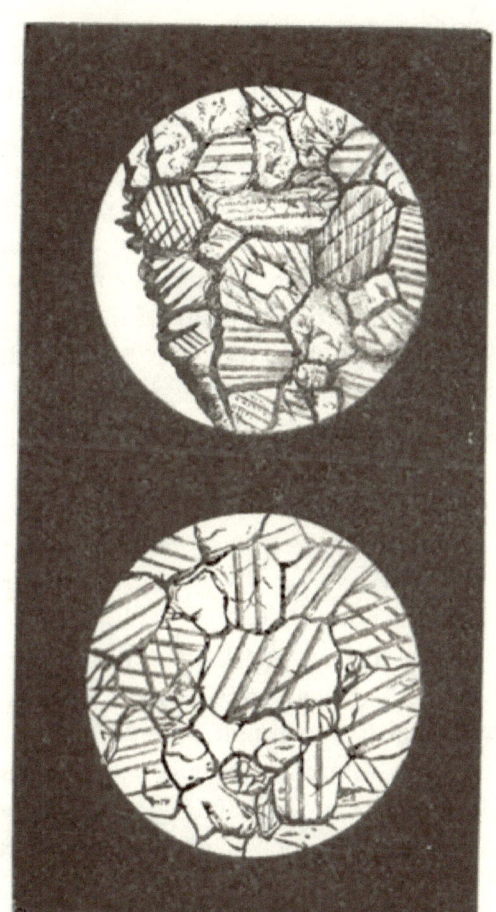

Fig. 24.—Microscopic structure of white marble employed in Edinburgh tombstones.

A, Structure of the fresh marble. B, Structure of the marble after standing eighty-seven years. The black edge is the crust of sulphate of lime and town dust which descends along rifts and cleavage planes.

mental rhombohedral cleavage is everywhere well developed. Not a trace exists of any amorphous granular matrix or base holding the crystalline grains together. These seem moulded into each other, but have evidently no extraordinary cohesion. A small fragment placed in dilute acid was entirely dissolved. There can be no doubt that this marble must be very nearly pure carbonate of lime.

The process of weathering in the case of this white marble presents three phases, sometimes to be observed on the same slab—viz. superficial solution, internal disintegration, and curvature with fracture.

(1.) *Superficial Solution* is effected by the carbonic acid, and partly by the sulphuric acid of town rain. When the marble is first erected it possesses a well-polished surface, capable of affording a distinct reflection of objects placed in front of it. Exposure for not more than a year or two to our prevalent westerly rains suffices to remove this polish, and to give the surface a rough granular character. The granules which have been cut across or bruised in the cutting and polishing process are first attacked and removed in solution, or drop out of the stone. An obelisk in Greyfriars Churchyard, erected in memory of a lady who died in 1864, has so rough and granular a surface that it might readily be taken for a sandstone. So loosely are the grains held together that a slight motion of the finger will rub them off. In the course of solution and removal, the internal structure of the marble begins to reveal itself. Its harder nests and veinings of calcite and other minerals project above the surrounding surface, and may be traced as prominent ribs and excrescences running across the faint or illegible inscriptions. On the other hand, some portions of the marble are more rapidly removed than others. Irregular channels, dependent partly on the direction given to trickling rain by the form of the monumental carving, but chiefly on original differences in the internal structure of the stone, are gradually hollowed out. In this way the former artificial surface of the marble disappears, and is changed into one that rather recalls the bare bleached rocks of some mountain-side.

The rate at which the transformation takes place seems to depend primarily on the extent to which the marble is exposed to rain. Slabs which have been placed facing to north-east, and with a sufficiently projecting architrave to keep off much of the rainfall, retain their inscriptions legible for a century or longer. But even in these cases the progress of internal disintegration is distinctly visible. Where the marble has been less screened from rain the rapidity of waste has been sometimes very marked. A good illustration is supplied by the tablet on the south side of Greyfriars Churchyard, erected in memory of G—— G——, who died in 1785.[1] This monument had become so far decayed as to require restoration in 1803. It is now, and has been for some years, for the most part utterly illegible. The marble has been dissolved away over the centre of the slab to a depth of about a quarter of an inch. Yet this monument is by no means in an exposed situation. It faces eastward in a rather sheltered corner, where, however, the wind eddies in such a way as to throw the rain against the part of the stone which has been most corroded.

In the majority of cases superficial solution has been retarded by the formation of a peculiar gray or begrimed crust, to be immediately described. The marble employed here for monumental slabs appears to be peculiarly liable to the development of this crust. Another kind of white

[1] For obvious reasons I withhold the names carved on the tomb stones here referred to.

marble, sometimes employed for sculptured ornaments on tombstones, dissolves without crust. It is snowy white and more translucent than the ordinary marble. So far as the few weathered specimens I have seen enable me to judge, it appears to be either Carrara marble, or one of the strongly saccharoid, somewhat translucent, varieties employed instead of it. This stone, however, though it forms no crust, suffers marked superficial solution. But it escapes the internal disintegration, which, so far as I have observed, is always an accompaniment of the crust. Yet the few examples of it I have met with hardly suffice for any comparison between the varieties.

(2.) *Internal Disintegration.* — Many of the marble monuments in our older churchyards are covered with a dirty crust, beneath which the stone is found on examination to be merely a loose crumbling sand, of incoherent calcite granules. This crust seems to form chiefly where superficial solution is feeble. It may be observed to crack into a polygonal network, the individual polygons occasionally curling up so as to reveal the yellowish white crumbling material underneath. It also rises in blisters which, when they break, expose the interior to rapid disintegration.

So long as this begrimed film lasts unbroken, the smooth face of the marble slab remains with apparently little modification. The inscription may be perfectly legible, and one would not readily believe the stone to be decayed at all. The moment the crust is broken up, however, the decay of the stone is rapid. For we then see that beneath the smooth, coherent surface-film the cohesion of the individual crystalline granules of the marble has already been destroyed, and that the merest touch causes them to crumble into a loose sand.

It appears, therefore, that two changes take place in upright marble slabs freely exposed to rain in our burial-grounds—a superficial, more or less firm crust is formed, and the cohesion of the particles beneath is destroyed.

The crust varies in colour from a dirty gray to a deep brown-black, and in thickness from that of writing-paper up to sometimes at least a millimetre. One of the most characteristic examples of it was obtained from an utterly decayed tomb (erected in the year 1792) on the east side of Canongate Churchyard. No one would suppose that the pieces of flat dark stone lying there on the sandstone plinth were once portions of white marble. Yet a mere touch suffices to break the black crust, and the stone at once crumbles to powder. Nevertheless the two opposite faces of the original polished slab have been preserved, and I even found the sharply-chiselled socket-hole of one of the retaining-nails. The specimen was carefully removed and soaked in a solution of gum, so as to preserve it from disintegration. On submitting the crust of this marble to microscopic investigation, I found it to consist of particles of coal and soot, grains of quartz-sand, angular pieces of broken glass, fragments of red brick or tile, and organic fibres. This miscellaneous collection of town dust was held together by some amorphous cement, which was not dissolved by hydrochloric acid. At my request my friend Mr. B. N. Peach tested it with soda on charcoal, and at once obtained a strong sulphur reaction. There can be little doubt that it is mainly sulphate of lime. The crust which forms upon our marble tombstones is thus a product of the reaction of the sulphuric acid of the town rain upon the calcium carbonate of the stone. A pellicle of amorphous gypsum is deposited upon the marble, and encloses the particles of dust which give the characteristic sooty aspect to the stone. This pellicle, when once formed, seems to be comparatively little affected by the chemical activity of rain-water. Hence the conservation of the even surface of the marble. It is liable, however, to be cracked by an internal expansion of the stone, to which I shall immediately refer, and also to rise in small blisters, and, as I have said, its rupture leads at once to the rapid disintegration of the stone.

The cause of this disintegration is the next point for consideration. Chemical examination revealed the presence of a slight amount of sulphate in the heart of the crumbling marble; but the quantity appeared to me to be too small seriously to affect the cohesion of the stone. I submitted to microscopic examination a portion of a crumbling urn of white marble in Canongate Churchyard. The tomb bears a perfectly fresh date of 1792 cut in sandstone over the top; but the marble portions are crumbling into sand, though the structure faces the east, and is protected from vertical rain by arching mason-work. A small portion of the marble retaining its crust was boiled in Canada balsam, and was then sliced at a right angle to its original polished surface. By this means a section of the crumbled marble was obtained, which could be compared with one of the perfectly fresh stone (see Fig. 24, B). From the dark outer amorphous crust, with its carbonaceous and other miscellaneous particles, fine rifts could be seen passing down between the separated calcite granules, which in many cases were quite isolated. The black crust descends into these rifts, and likewise passes along the cleavage planes of the granules. Towards the outer surface of the stone, immediately beneath the crust, the fissures are chiefly filled with a yellowish, structureless substance, which gave a feeble glimmering reaction with polarised light, and enclosed minute amorphous aggregates like portions of the crust. It probably consists chiefly of sulphate of lime. But the most remarkable feature in the slide was the way in which the calcite granules had been corroded. Seen with reflected light they resembled those surfaces of spar which have been placed in weak hydrochloric acid to lay bare enclosed crystals of zeolites. The solution had taken place partly along the outer surfaces, so as to produce the fine passages or rifts, and partly along the cleavage. Deep cavities, defined by intersecting cleavage planes, appeared to descend into the heart of some of the granules. In no case did I observe any white pellicle such as might indicate a re-deposit of lime from the dissolved carbonate. Except for the veinings of probable sulphate just referred to, the lime, when once dissolved, had apparently been wholly removed in solution. There was further to be observed a certain dirtiness, so to speak, which at the first glance distinguished the section of crumbled marble from the fresh stone. This was due partly to corrosion, but chiefly to the introduction of particles of soot and dust, which could be traced among the interstices and cleavage lamellæ of the crystalline granules for some distance back from the crust.

It may be inferred, therefore, that the disintegration of the marble is mainly due to the action of carbonic acid in the permeating rain-water, whereby the component crystalline granules of the stone are partially dissolved and their mutual adhesion is destroyed. This process goes on in all exposures and with every variety in the thickness of the outer crust. It is distinctly traceable in tombstones that have not been erected for more than twenty years. In those which have been standing for a century it is, save in exceptionally sheltered positions, so far advanced that a very slight pressure suffices to crumble the stone into powder. But with this internal disintegration we have to take into consideration the third phase of weathering to which I have alluded. In the upright marble slabs it is the union of the two kinds of decay that leads to so rapid an effacement of the monuments.

(3.) *Curvature and Fracture.*—This most remarkable phase of rock-weathering is only to be observed in the slabs of marble which have been firmly inserted into a solid framework of sandstone, and placed in an erect or horizontal position. It consists in the bulging out of the marble accompanied with a series of fractures. This change cannot be explained as mere sagging by gravitation, for it usually appears as a swelling up of the centre of the slab, which continues until the large blister-like expansion is ruptured. Nor is it by any means exceptional; it occurs, as a rule, on all the older upright marble tablets, and is only found to be wanting in those cases where the marble has evidently not been fitted tightly into its sandstone frame. Wherever there has been little or no room for expansion, protuberance of the marble may be observed. Successive stages may be seen from the first gentle uprise to an unsightly swelling of the whole stone. This change is accompanied by fracture of the marble. The rents in some cases proceed from the margin inwards, more particularly from the upper and under edges of the stone, pointing unmistakably to an increase in volume as the cause of fracture. In other cases the rents appear in the central part of the swelling where the tension from curvature has been greatest.

Some exceedingly interesting examples of this singular process of weathering are to be seen in Greyfriars Churchyard. On the south wall, in the enclosure of a well-known county family, there is an oblong upright marble slab facing west, and measuring $30\frac{1}{4}$ inches in height by $22\frac{3}{8}$ inches in breadth and $\frac{3}{4}$ inch in thickness. The last inscription on it bears the date 1838, at which time, of course, it was no doubt still smooth and upright. Since then, however, it has escaped from its fastenings on either side, though still held firmly at the top and bottom. It consequently projects from the wall like a well-filled sail. The axis of curvature is, of course, parallel to the upper and lower margins, and the amount of deviation from the original vertical line is fully $2\frac{1}{2}$ inches, so that the hand and arm can be inserted between the curved marble and the perfectly vertical and undisturbed wall to which it was fixed. At the lower end of the slab a minor curvature, to the extent of $\frac{1}{8}$ of an inch, is observable, coincident with the longer axis of the stone. In both cases the direction of the bending has been determined by the position of the enclosing solid frame of sandstone which resisted the internal expansion of the marble. Freed from its fastenings at either side the stone has assumed a simple wave-like curve. But the tension has become so great that a series of rents has appeared along the crest of the fold. One of these has a breadth of $\frac{1}{16}$ of an inch at its opening.[1] Not only has the slab been ruptured, but its crust has likewise yielded to the strain, and has broken up into a network of cracks, and some of the isolated portions are beginning to curl up at the edges, exposing the crumbling decayed marble below. I should add that such has been the expansive force of the marble that the part of the sandstone block in the upper part of the frame, exposed to the direct

---

[1] It is a further curious fact that the slab measures $\frac{1}{2}$ inch more in breadth across the centre where it has had room to expand, than at the top, where it has been tightly jammed between the sandstone slabs.

pressure, has begun to exfoliate, though elsewhere the stone is quite sound.

More advanced stages of curvature and fracture may be noticed on many other tombstones in the same burying-place. One of the most conspicuous of these has a peculiar interest from the fact that it occurs on the tablet erected to the memory of one of the most illustrious dead whose dust lies within the precincts of the Greyfriars—the great Joseph Black. He died in 1799. In the centre of the sumptuous tomb raised over his grave is inserted a large upright slab of white marble, which, facing south, is protected from the weather partly by heavy overhanging masonry and partly by a high stone wall immediately to the west. On this slab a Latin inscription records with pious reverence the genius and achievements of the discoverer of carbonic acid and latent heat; and adds, that his friends wished to mark his resting-place by the marble whilst it should last. Less than eighty years, however, have sufficed to render the inscription already partly illegible. The stone, still firmly held all round its margin, has bulged out considerably in the centre, and the blister-like expansion has been rent by numerous cracks, which run, on the whole, in the direction of the length of the stone.

A further stage of decay is exhibited by a remarkable tomb on the west wall of the Greyfriars Churchyard. The marble slab, bearing a now almost wholly effaced inscription, on which the date 1779 can be seen, is still held tightly within its enclosing frame of sandstone slabs, which are firmly built into the wall. But it has swollen out into a ghastly protuberance in the centre, and is, moreover, seamed with rents which strike inwards from the margins. In this and in some other examples the marble seems to have undergone most change on the top of the swelling, partly from the system of fine fissures by which it is broken up, and partly from more direct and effective access of rain. Eventually the cohesion of the stone at that part is destroyed, and the crumbling marble falls out, leaving a hole in the middle of the slab. When this takes place, disintegration proceeds rapidly. Three years ago I sketched a tomb in this stage on the east wall of Canongate Churchyard. In a recent visit to the place I found that the whole of the marble had since fallen out.

The first cause that naturally suggests itself in explanation of the remarkable change in the structure of a substance usually believed to be so inelastic as white marble, is the action of frost. White statuary marble is naturally porous. It is rendered still more so by that internal solution which I have described. The marble tombstones in our graveyards are therefore capable of imbibing a relatively large amount of moisture. When this insterstitial water is frozen, its expansive force, as it passes into the solid state, must increase the isolation of the granules and augment the dimensions of a marble block. I am inclined to believe that this must be the principal cause of the change. Whatever may be the nature of the process it is evidently one which acts from within the marble itself. Microscopic examination fails to discover any chemical transformation which would account for the expansion. Dr. Angus Smith has pointed out that in towns the mortar of walls may be observed to swell up and lose cohesion from a conversion of its lime into the condition of sulphate. I have already mentioned that sulphate does exist within the substance of the marble, but that its quantity, so far as I have observed, is too small to be taken into account in this question. The expansive power is exerted in such a way as not sensibly to affect the internal structure and composition of the stone. And this I imagine is most probably the work of frost.

The results of my observations among our burial-grounds show that, save in exceptionally sheltered situations, slabs of marble, exposed to the weather in such a climate and atmosphere as that of Edinburgh, are entirely destroyed in less than a century. Where this destruction takes place by simple comparatively rapid superficial solution and removal of the stone, the rate of lowering of the surface amounts sometimes to about a third of an inch (or roughly 9 millimetres) in a century. Where it is effected by internal displacement, a curvature of $2\frac{1}{2}$ inches, with abundant rents, a partial effacement of the inscription, and a reduction of the marble to a pulverulent condition, may be produced in about forty years, and a total disruption and effacement of the stone within one hundred. It is evident that white marble is here utterly unsuited for out-of-door use, and that its employment for works of art which are meant to stand in the open air in such a climate ought to be strenuously resisted. Of course I am now referring not to the durability of marble generally, but to its behaviour in a large town with a moist climate and plenty of coal-smoke.

II. SANDSTONES AND FLAGSTONES.—These, being the common building materials of the country, are of most frequent occurrence as monumental stones, and where properly selected are remarkably durable. By far the best varieties are those which consist of a nearly pure fine siliceous sand, with little or no iron or lime, and without trace of bedding or other structure. Some of them contain as much as 98 per cent of silica. A good illustration of their power of resisting the weather is supplied by Alexander Henderson's tomb in Greyfriars Churchyard. He died in 1646, and a few years afterwards the present tombstone, in the form of a solid square block of freestone, was erected over his grave. It was ordered to be defaced in 1662 by command of the Scottish Parliament, but after 1688 it was repaired. Certain bullet marks upon the stone are pointed out as those of the soldiery sent to execute the order. Be this as it may, the original chisel marks on the polished surface of the stone are still perfectly distinct, and the inscribed lettering remains quite sharp. Two hundred years have effected hardly any change upon the stone, save that on the west and north sides, which are those most exposed to wind and rain, the surface is somewhat roughened, and the internal fine parallel jointing begins to show itself.

Three obvious causes of decay in arenaceous rocks may be traced among our monuments. In the first place, the presence of a soluble or easily removable matrix in which the sand-grains are embedded. The most common kinds of matrix are clay, carbonates of lime and iron, and the anhydrous and hydrous peroxides of iron. The presence of the iron reveals itself by its yellow, brown, or red colour. So rapid is disintegration from removal of the matrix that the sharply-incised date of a monument erected in Greyfriars Churchyard to an officer who died only in 1863 is no longer legible. At least $\frac{1}{8}$ of an inch of surface has here been removed from a portion of the slab in sixteen years, or at the rate of about three-quarters of an inch in a century.

In the second place, where a sandstone is marked by distinct laminæ of stratification, it is nearly certain to split up along these planes under the action of the weather, if the surface of the bedding-planes is directly exposed. This is well known to builders, who are quite aware of the importance of "laying a stone on its bed." Examples may be observed in our churchyards where sandstones of this character have been used for pilasters and ornamental work, and where the stone, set on its edge, has peeled off in successive layers. In flagstones, which are merely thinly-bedded sandstones, this minute lamination is often fatal to durability. These stones, from the large size in which slabs of them can be obtained, and from the ease with which they can be worked, form a tempting material for monumental inscriptions. The melancholy result of trusting to their permanence is strikingly shown by a tombstone at the end of the south burying-ground in Greyfriars Churchyard. The date inscribed on it is 1841, and the lettering that remains is as sharp as if cut only recently. The stone weathers very little by surface disintegration. It is a laminated flagstone set on edge, and large portions have scaled off, leaving a rough, raw surface where the inscription once ran. In this instance a thickness of about $\frac{1}{8}$ of an inch has been removed in forty years.

In the third place, where a sandstone contains concretionary masses of different composition or texture from the main portion of the stone, these are apt to weather at a different rate. Sometimes they resist destruction better than the surrounding sandstone so as to be left as permanent excrescences. More commonly they present less resistance, and are therefore hollowed out into irregular and often exceedingly fantastic shapes. Examples of this kind of weathering abound in our neighbourhood. Perhaps the most curious to which a date can be assigned are to be found in the two sandstone pillars which until recently flanked the tomb of Principal Carstares in Greyfriars Churchyard. They were erected some time after the year 1715. Each of them is formed of a single block of stone about 8 feet long. 'Exposure to the air for about 150 years has allowed the original differences of texture or composition to make their influence apparent. Each column is hollowed out for almost its entire length on the exposed side into a trough 4 to 6 inches deep and 6 to 8 inches broad. As they lean against the wall, beneath the new pillars which have supplanted them, they suggest some rude form of canoe rather than portions of a sepulchral monument.

Where concretions are of a pyritous kind their decomposition gives rise to sulphuric acid, some of which combines with the iron and gives rise to dark stains upon the corroded surface of the stone. Some of the sandstones of the district, full of such impurities, ought never to be employed for architectural purposes. Every block of stone in which they occur should be unhesitatingly condemned. Want of attention to this obvious rule has led to the unsightly disfigurement of public buildings.

III. GRANITES.—In Professor Pfaff's experiments, to which I have already referred, he employed plates of syenite and granite, both rough and polished. He found that they had all lost slightly in weight at the end of a year. The annual rate of loss was estimated by him as equal to 0·0076 mm. from the unpolished, and 0·0085 from the polished granite. That a polished surface of granite should weather more rapidly than a rough one is perhaps hardly what might have been expected. The same observer remarks that though the polished surface of syenite was still bright at the end of not more than three years, it was less so than at first; and, in particular, that some figures indicating the date, which he had written on it with a diamond, had become entirely effaced. Granite has been employed for too short a time as a monumental stone in our cemeteries to afford any ready means of measuring even approximately its rate of weathering. Traces of decay in some of its felspar crystals may be detected, yet in no case that I have seen is the decay of a polished granite surface sensibly apparent after exposure for fifteen or twenty years. That the polish will disappear, and that the surface will gradually roughen as the individual component crystals are more or less easily attacked by the weather, is of course sufficiently evident. Even the most durable granite will probably be far surpassed in permanence by the best of our siliceous sandstones. But as yet the data do not exist for making any satisfactory comparison between them.

Since the preceding pages were written, I have had an opportunity of examining the condition of the monumental stones in the graveyards of a number of towns and villages in the north-east of Scotland, where the population is sparse and where comparatively little coal-smoke passes into the atmosphere. The marble tablets last longer there than in Edinburgh, but show everywhere indications of decay. They appear to be quite free from the black or gray sulphate-crust. They suffer chiefly from superficial erosion, but I observed a few cases of curvature and fracture. As a contrast to the universal decay of the marble tombstones, reference may be made to the remarkable durability of the clay-slate which has been employed for monumental purposes in Aberdeenshire. It is a fine-grained, rather soft rock, containing scattered cubes of pyrites, and capable of being readily dressed into thin smooth slabs. A tombstone of this material, erected in the old burying-ground at Peterhead, sometime between 1785 and 1790, retains its lettering as sharp and smooth as if only recently incised. Yet the stone is soft enough to be easily cut with the knife. The cubes of pyrites have resisted weathering so well that a mere thin film of brown hydrous peroxide conceals the brassy undecomposed sulphide from view. The slate is slightly stained yellow round each cube or kernel of pyrites, but its general smooth surface is not affected. The lapse of nearly a century has produced scarcely any change upon this stone, while neighbouring tablets of white marble, 100 to 150 years old, present rough granular surfaces and half-effaced though still legible inscriptions.

IX.

IN WYOMING.[1]

Twenty-four hours after landing in New York my preparations for a journey to the Far West were completed, and I found myself looking out from the windows of a Pullman car that rapidly swept past the blue reaches of the Hudson. A project which had been little more than a dream for many years was now at last actually realised. Let me briefly explain this project, that the purport of the journey, and of the following notes, may be understood.

And first I would give the reader due warning that the object of the expedition was not sport or adventure, but science. My companion and I were not, indeed, wholly unarmed. To go without at least revolvers into these western wildernesses would, we were told, be sheer folly. My weapon disappeared, however, in an early part of our travels, but my friend's did occasional service upon a badger or prairie hen. All the sport that was done consisted in the slaughter of the antelope or elk that was needed for food. Nevertheless, from first to last, the journey was full of interest, and, in a quiet way, even of excitement. We had game of our own to hunt, and we pursued it with such measure of success as at least amply to justify our own expectations, and to reward us for the enterprise.

[1] *Macmillan's Magazine*, 1881.

Everybody now knows the kind of evidence from which it has been established that the present surface of the dry land has once been in a wholly different condition. In all parts of the world this evidence obtrudes itself, often so conspicuously as from earliest times to have arrested the attention of mankind, and to have suggested, or at least coloured, mythology and local superstition. In many places, for example, as soon as the layer of soil or subsoil has been removed, the rock below, with its embedded shells or corals, or other remains of marine life, is at once seen to have been the bottom of the sea. At other points we find traces of rivers which must have had their sources in mountains that have long since disappeared, and which fed lakes or watered woodlands and plains that for ages have been buried out of sight. Or, again, we come upon the earth and stones left by vanished glaciers, upon the limestone spread out by springs long ago dried up, upon the sheets of lava or heaps of ashes thrown out by volcanoes that have been extinct and effaced for ages. It is manifest, therefore, that the present surface of the land, so far from being aboriginal, is only the latest phase of a long succession of geographical revolutions, the uppermost leaf as it were, of a series of volumes that lie beneath it. Mountains and hills, valleys and plains, instead of standing out as parts of the primeval architecture of the globe, can be shown to belong to many different epochs of the earth's long history.

But the question remains, how these familiar features have come to be impressed on the surface of the land. Granted that the solid materials out of which a mountain or tableland has been built were originally accumulated as sediment on the floor of the sea, how has this hardened sediment been fashioned into the well-known lineaments of the land? The solution of this question aroused some years ago a keen discussion, and has given rise to a portentous mass of geological literature. The combatants, as in most warfares, scientific or other, ranged themselves into two camps. There were the Convulsionists, or believers in the paramount efficacy of subterranean movement, who, starting from the universally admitted proofs of upheaval, crumpling, and fracture, sought an explanation of the present inequalities of the land in unequal disturbance from below. On the other hand, there were the Erosionists, or upholders of the efficacy of superficial waste, who maintained that besides the elevations due to subterranean causes, mountains, valleys, and all the other features of a landscape, have been gradually carved into their present shapes by the slow abrasion of the air, rain, rivers, frosts, and the other agents of subaërial erosion. The contest, which was keen enough some years ago, has for a while almost ceased among us, though an occasional shot from younger combatants, fired with the old enthusiasm, serves to keep alive the memory of the campaign.

Having long ago attached myself to the camp of the Erosionists, though by no means inclined to do battle under the extreme "quietest" banners of some of its champions, I have been led, in the course of my wanderings over this country and the Continent, to look at scenery with a peculiar interest. I have long been convinced, however, that for the proper discussion of the real efficacy of superficial erosion in the development of a terrestrial surface, the geologists of Europe have been at great disadvantage. The rocks in these regions have undoubtedly been subjected to so many changes—squeezed, crumpled, fractured, upheaved, and depressed—that the effects of unequal erosion upon their surface have been masked by those of subterranean disturbance. The problem has thus become much more complicated than with simpler geological structure it would have been. Its solution has demanded an amount of knowledge of geological structure which can hardly be acquired without long and laborious training, the want of which on the part of many who have taken part in the controversy has led to the calling in question or denial of facts, about the reality and meaning of which there should never have been any doubt at all. That, in spite of these obstacles, observers in this country should have been able to brush aside the accidental or adventitious difficulties and get at the real gist of the matter, as I am certain they have done, seems to me a lasting proof of their scientific prowess.

Now, it is unquestionably true that had the birthplace of geology lain on the west side of the Rocky Mountains, this controversy would never have arisen. The efficacy of denudation instead of evoking doubt, discussion, or denial, would have been one of the first obvious principles of the science, established on the most irrefragable basis of patent and most impressive facts. Over thousands of square miles in that region the strata remain practically unchanged from their original horizontal position, so that the effects of surface erosion can at once be detected upon their flat parallel layers. The country has not been under the sea for a vast succession of geological periods. It has not been buried, like so much of Northern Europe and North-Eastern America, under a thick cover of ice-borne clays and gravels. Its level platforms of sandstone, shale, clay, or limestone lie at the surface, bare to the wind and rain, and their lines can be followed mile after mile, as if the whole region were one vast geological model to which the world should come to learn the fundamental laws of denudation.

For the exploration of these western territories the

enlightened enterprise of various departments under the American Government has already done a great deal. During the last ten or fifteen years various surveys of different portions of the region have been carried on, and a voluminous series of maps and reports has been issued embodying the results of the explorations. Through the courteous liberality of these departments, for which on all occasions I am anxious to express my gratitude and admiration, I had received copies of most of their publications. The descriptions of King, Hayden, Powell, Gilbert, Dutton, Emmons, Hague, Marvine, Endlich and others, and the remarkable drawings of Holmes, had made me in some respects familiar with the general aspects of the scenery and geological structure of the region. From these works it was evident that questions over which we had been fighting so long in Europe were finally settled by Nature herself in America, after a fashion admitting of no more cavil. It was well worth while to make a journey to the far West to see with one's own eyes the demonstration for which one had longed on this side of the Atlantic. And this was what I now had determined to do, with the companionship of an old friend of kindred tastes, Mr. Henry Drummond, of the New College, Glasgow, who from first to last shared in the work and smoothed the little privations of the journey.

Of the travelling westward, now made so familiar and comparatively easy by the various rival railroad companies, little need be said here. There is an early and late feature of it, however, to which reference may be made, partly in the hope that every renewed protest against an abuse, as offensive to many of our cousins on the other side as to a visitor from the old country, may help towards its ultimate suppression. Hardly is the traveller out of New York than he notices that every natural rock, islet, or surface of any kind that will hold paint is disfigured with advertisements in huge letters. The ice-worn bosses of gneiss which rising out of the Hudson, would in themselves be such attractive objects in the landscape, are rendered hideous by being made the groundwork on which some kind of tobacco, or tooth-wash, or stove-polish, is recommended to the notice of the multitude. All the way west to the Pacific along the railway route the same barbarous practice has been employed, with an ingenuity and perseverance worthy of a better cause. Some of the most picturesque cañons on the route have had their walls turned into advertising boards —for the spoilers have travelled with ladders as well as paint-pots, and have carefully inscribed their wares on precipices which would ordinarily be inaccessible. Oil-paint lasts for many years; so that even if the sacrilege be soon suppressed it will be long before the record of it has wholly disappeared.

Not many years ago Chicago lay at the extreme verge of advancing civilisation. One who had been so far west could boast that he had reached the limits of settlements, and had looked on the great plains haunted by wild red men and buffaloes. Now, however, the network of railways has spread far beyond Chicago, which has become one of the chief marts of the Union, having free communication alike by water and land with the eastern seaboard of the continent. I was making some such natural reflections as the train slowed in approaching Chicago station, when a noise as of broken glass came from the other unoccupied end of the car. The crash was loud enough to startle everybody for a moment, but the conversation and packing up of bags were immediately resumed. On going to the spot I found that two window-panes of the car had been pierced at about the same height by two successive bullets from a revolver. One of the balls had made a clean sharp hole in the plate-glass, and would no doubt have continued its journey through the body of any unfortunate occupant of the seat. This was our first experience of "Western Life." We looked next morning in the papers for an account of the "outrage," as it would have been termed by our penny-a-liners at home. It was not mentioned at all. We found, however, records of so many successful shootings that the non-insertion of our episode was easily to be explained. The incident impressed me with a sense of recklessness in the use of firearms and disregard of life—an impression that was not effaced by the rest of the journey.

We crossed the Mississippi at night, and having some time to wait at the Quincy Junction walked down to the banks of the river and reverently dipped our hands in the great "Father of Waters." Lights gleamed from the farther side, heightening the effect of vastness and mystery. Behind us, too, gleamed the much brighter lights of rival drinking saloons, from which, before resuming the journey, we were enabled to enlarge our rapidly-growing vocabulary of American drinks.

The Missouri River at Kansas City is the muddiest, most tumultuous flood of rolling water I ever saw. Yet it was now the month of August, and there had been a long course of previous dry weather. The train carried us slowly across a creaking wooden bridge over the boiling sea below, past some cliffs of old alluvium, into a station full of negroes, of whom there had been a large influx from the South in search of a proposed settlement in Kansas. There being now some kind of picnic or holiday afoot, they were a merry, noisy crowd, dressed out and bedizened as only niggers can be. One seldom realises what an extraordinary variety of tint there may be in a coloured population. Some of the excursionists were of the most perfect coal-black shade, from which every gradation could be noted till the crisp hair and characteristic features remained as almost the only traces of negro blood. Westward still, through endless monotonous miles of maize and yet unbroken land, the train moved wearily hour after hour, until on getting up in the morning we found ourselves unmistakably on the great prairie at last. Perhaps no type of scenery so closely fulfils a previous mental picture of it as the western prairie of North America. Seen after a hot summer, it spreads out as a vast, treeless, arid expanse, covered with a short and sparse grass, which, though green and flowery in spring, becomes parched by drought into a kind of hay, through which the baked soil everywhere peeps. For hundreds of miles together the undulations never rise into hills nor sink into valleys. A sluggish streamlet, depressed a few feet or yards beneath the general level, winds here and there in lazy curves till it joins some sluggish and muddy tributary of the Missouri, that creeps along a level plain bounded by low bluffs. But ere autumn comes many of these watercourses have been reduced to groups of stagnant pools.

At proper intervals stations have been built with means for supplying the engines with water and fuel. It was at one of these halting-places that we were able to set foot for the first time on the prairie. The brief halt enabled us to make some observations that served materially to beguile the tedium of this railway journey, and to invest the featureless prairie with a new interest. Every traveller across the continent has remarked the incredible number of ant-hills and burrows of the prairie dog and gopher by which the flat bare surface of the plains is marked. The ground appears to be in a constant state of cutaneous eruption. So leisurely does the train move along, however, and so seldom does it halt, that for some hours after daylight we sat looking on this singular scene before an opportunity came of getting down to have a closer view of it. We noticed that though the general colour of the soil is a dirty yellowish gray or drab, the ant-hills have a somewhat pinkish tint. Our first halt revealed the curious fact that this difference arises from the choice which the ants make of their building materials. With infinite labour they pick up from the surface of the prairie the small broken crystals of flesh-coloured felspar that are sparsely strewed there. The rocks underneath are various sandstones, clays, and limestones, the decomposition of which could never have furnished this felspar detritus. I examined a good many ant-hills, and found the same kind of fragments on all of them. The felspar grains were most abundant, but there occurred also small pieces of quartz and other minerals of crystalline rocks, and here and there some black glistening specks of coal. There seemed to be a thin crust or veneering of this kind of fine detritus over the drab-tinted soil, not thick enough to be readily observable, but yet sufficiently persistent to supply the materials so patiently gathered together into these little mounds.

No warning bell gives the traveller notice to resume his place in the cars, and we had just time after hearing the "All aboard!" of the conductor to regain the train, more puzzled than ever by the prairie ant-hills. The source of this fine felspar drift, and the cause of its being spread so thinly over the many hundreds of square miles it evidently covered, were questions in the history of the prairies which we could not answer, but to which we were able to return with more light and increased interest on the homeward journey. At last, on the far western horizon the first summits of the Rocky Mountains rose like blue islets out of the sea. Hour after hour, as the train ground its dusty way over the plain, these islets rose higher, till at last they united into the long noble range of the snow-streaked Colorado Alps, with Pike's Peak, Long's Peak, and a host of other broad-based cones towering far up into the clear air.

Though it was no part of our programme to linger among these mountains, we gladly availed ourselves of the opportunity of making an excursion into them in passing. The first few hours showed us on what a different plan these mountains had been constructed from that which is more familiar in the Old World. Approaching the Alps, for instance, you cross a succession of parallel minor ranges, or foot-hills, like the Jura, which flank the more colossal ramparts behind them. But these Colorado Mountains tower straight out of the plain. For hundreds of miles to the east the Cretaceous or Tertiary strata underlying the prairie seem to be nearly flat or only very slightly undulating, though there is a steady rise of the ground westward. But at the foot of the mountains they are at once abruptly pitched up on end. So sharp and sudden is the bend that it would hardly be an exaggeration to say that you might sit on the flat beds and lean your back on the vertical ones. From some points of view the solid sheets of rock made a magnificent curve from the plains up into the line of serrated crags which their broken edges present against the sky. The meaning of this structure is soon apparent when the traveller ascends one of the numerous deep gorges or cañons into which the flanks of the mountains have been trenched by the erosion of the escaping drainage. In the course of a brief space he finds that he has crossed the uptilted formations and has reached the ancient granitic and crystalline rocks, which have been driven up like a huge wedge through the younger strata of the prairies, and now form the axis of the Colorado Mountains. But for the protrusion of this wedge the "Centennial State" would have been a quiet pastoral or agricultural territory like the region to the eastward. The rise of the granitic axis, however, has brought up with it that incredible mineral wealth which, in a few years, has converted the loneliest mountain solitudes into busy hives of industry. Places that a few years ago were haunted only by wild beasts, and probably hardly ever saw even a red man, now count their population by thousands. Mining camps have grown into cities with important public buildings, hotels, and many of the luxuries as well as vices of modern city life. There is a feverish rush westward. Advertisements placarded all over the Union by rival railroad companies show the cheapest and quickest route to the new El Dorado of Colorado, and hold out tempting prospects of rapidly acquiring a fortune there. We found ourselves unwittingly moving westward on this wave of emigration. It was tacitly assumed that we too were bound for a "claim" somewhere.

After a glimpse at the cañons and camp-life of these uplands, we skirted their eastern slopes amid mounds of *débris*, which renewed our interest in the problem that had been started by the prairie ant-hills. Without halting at that time, however, but pursuing our way westward by the Union Pacific Railroad, we made no stop till we came within sight of the Uintah Mountains in Wyoming. This long journey is marked in the recollection of a traveller by the complete demolition of his previous mental picture of the "Rocky Mountains." Misled by the absurd and utterly false system, still far from extinct, of representing a watershed on a map by a continuous range of mountain chain, most people have grown up in the belief that the backbone of North America consists of a colossal rampart of mountains which traverses the continent as a continuous range, running in a nearly north and south direction, and so extraordinarily rugged as to have deserved the special appellation of "Rocky." No conception could well be further from the reality. To depict the American watershed in this way is nearly as erroneous as it would be to draw a lofty mountain chain from the Pyrenees across the heart of France, Switzerland, Germany, and Russia, as indicative of the watershed of Europe. Such is the force of habit engendered by the long use of faulty maps that though we knew what the true structure of the country had been shown to be, it was yet with a feeling almost of incredulity that we looked out upon the scene on either side of the railroad track as the train approached the summit of the route. The Colorado Alps had sunk down into a series of low ridges, though we could still see in the far distance some of their more notable peaks. Northward the tops of some distant hills in Wyoming loomed up on the horizon, but all around us not only were there no mountains, but hardly anything that deserved to be called a hill—certainly nothing that for a moment suggested the crest of a mountain range. The railway company, with a laudable desire for the diffusion of correct geographical knowledge, has had a board inscribed "Summit of the Rocky Mountains," and placed at the highest level of their line. One looks round with a feeling of disappointment for the peaks and crests that ought to have been there. Instead of these, there is the same long smooth prairie-like slope, out of which rise numerous quaint knobs of pink granite. The central wedge not having been driven so far upward here forms no conspicuous feature at the surface. Yet it has carried up the same red sandstones on its eastern flank that rise in vertical bands among the cañons north of Denver. From the plain of the Missouri the prairie, there about 1000 feet above sea-level, rises slowly in elevation westward, till at Cheyenne, a distance of rather more than 500 miles, its surface has an average elevation of about 6000 feet. In the next eighteen miles, however, it makes a more rapid slope, for it mounts to an elevation of 8271 feet above the sea. The loss of the cherished delusion about the aspect of the Rocky Mountains was in some small measure compensated by a glimpse we had of the source whence the prairies have derived their fine detritus and the ants their favourite pink building materials. The granite of this elevated plateau is a bright flesh-coloured rock crumbling into sand, the grains of which are mainly of pink cleavable orthoclase felspar. Exposed to all the vicissitudes of weather at so great an altitude, the rock readily disintegrates. Every shower of rain washes down some of its detritus, which is further carried far over the plains by wind. It was no doubt from such a rock as this that the widespread felspar drift of the prairie has been derived, and this very ridge has probably furnished a due amount of it.

After crossing the summit, the railroad track descends slowly into the elevated plateau known as the Laramie Plains, which still drain eastward into the Atlantic. Not until the train has crossed this dreary region for some 150 miles or more, does it reach the true watershed of the country. And then, instead of a colossal rampart of rugged mountains, we find still the same monotonous plains on which the few names that have been affixed to localities—Red Desert, Bitter Creek, Salt Wells, and others — sufficiently denote the sterile character of the region. We were now among the head-waters of the great Colorado River on the Pacific slope of the continent. But of visible slope there is for a long way no trace. It is a bare, treeless, verdureless waste, crumbling under the fierce glare of a cloudless sky and the hot blast of a parching wind. Yet for long ages these deserts were the site of a succession of lakes vaster in size than any now existing on the American continent. The water has disappeared, and out of the hardened clay and marl of the lake bottoms the elements are carving some of the weirdest scenery on the face of the earth. Every mile of the dusty journey now brought with it new and still stranger proofs of this marvellous erosion. At one moment we were looking out on what might have been taken for the bastions of a fort that had stood a long siege. Another curve of the line brought into view seemingly the mouldering battlements and decayed acropolis of some early heroic city; at the next turn the array of rock-forms could find no adequate parallel in human architecture. Scenery more indescribable can hardly be conceived. As yet, indeed, all we could see or know of these "Bad Lands" was from the windows of the car. But we saw clearly enough by their level lines of stratification that their forms had been sculptured out of horizontal rocks by surface agents. League after league this lesson of utterly inconceivable waste rose out impressively on either side, until at last, when we reached Carter Station, we almost felt that we had seen about as much as our faculties could very well assimilate. But much more was in store for us.

Thanks to the thoughtful kindness of my friends Dr. F. V. Hayden, to whom the geology of Western America owes so much, and Dr. Joseph Leidy, the revered Nestor of American comparative anatomy, Judge Carter was waiting our arrival, and soon carried us off, bag and baggage, to his hospitable home at Fort Bridger. In former days, before railway communication was opened across the continent, Fort Bridger was an important station on the emigrant road to Salt Lake and the Pacific Coast. It is now no longer a military post, and being at a distance from the present highway of traffic, some of its disused buildings are falling into disrepair. But Judge Carter, who used to be the patriarch of the district, still lives at his post, combining in his own worthy person the offices of postmaster, merchant, farmer, cattle-owner, judge, and general benefactor of all who claim his hospitality. His well-known probity has gained him the respect and goodwill of white man and red man alike, and we found his name a kind of household word all through the West. So rapidly and completely have things been changed on this route by the formation of the railway, that in listening to Judge Carter's stories of the olden time one could hardly realise that some of the most startling of them did not go further back than fifteen or twenty years. Horse-stealing would appear to have been the one unpardonable sin in these quarters. You might kill a man outright, and it might be nobody's affair either to avenge him or to see you brought to justice for the murder. But to steal his horse was to leave him to perish on the plains; and if you stole his horse this week you might return and steal mine next. So the best method of preventing that mishap was to put it out of your power ever to steal again. Killing you was consequently not murder; it was merely punishing effectually an offence that could not be reached by any ordinary legal means, in a region where criminals were many and police were none. Judge Carter had had many experiences of horse-stealers. On one occasion, travelling eastward across the prairie with his wife and family, he found next morning the horses stolen. Such a position resembles that of a ship at sea without masts or sails. There was no station at which provisions could be procured, so that the loss of the means of transport meant starvation and death. Fortunately the Judge succeeded in recovering his animals. On another occasion, having tried and convicted a horse-stealer, he sent him in custody to the court in Utah. The man was chained hands and feet, and in the course of the journey succeeded in breaking his foot-chain, and though still manacled, tried to escape. He was of course speedily shot by the two men who had been entrusted with the mission, and who were probably a couple of dare-devils no whit better than himself. They consulted as to their next step, and finding in their writ that they were "to deliver the body of the prisoner" to the sheriff at Salt Lake City, they took the instructions in their literal sense, stowed the body into the stage-coach, and delivered it duly at its destination.

From Fort Bridger the Judge carried us to see the "Mauvaises Terres," or "Bad Lands" of Wyoming. This expressive name has been given to some of the strangest and, in many respects, most repulsive scenery in the world. They are tracts of irreclaimable barrenness, blasted and left for ever lifeless and hideous. To understand their peculiar features it is needful to bear in mind that they lie on the sites of some of the old lakes already referred to, and that they have been carved out of flat sheets of sandstone, clay, marl, or limestone, that accumulated on the floors of these lakes. Everywhere, therefore, horizontal lines of stratification meet the eye, giving alternate stripes of buff, yellow, white, or red, with here and there a strange verdigris-like green. These strata extend nearly horizontally for hundreds of square miles. But they have been most unequally eroded. Here and there isolated flat-topped eminences or "buttes," as they are styled in the West, rise from the plain in front of a line of bluff or cliff to a height of several hundred feet. On examination, each of these hills is found to be built up of horizontal strata, and the same beds reappear in lines of terraced cliff along the margin of the plain. A butte is only a remnant of the original deep mass of horizontal strata that once stretched far across the plain. Its sides and the fronts of the terraced cliffs, utterly verdureless and bare, have been scarped into recesses and projecting buttresses. These have been further cut down into a labyrinth of peaks and columns, clefts and ravines, now strangely monumental, now uncouthly irregular, till the eye grows weary with the endless variety and novelty of the forms. Yet beneath all this chaos of outlines there can be traced everywhere the level parallel bars of the strata. The same band of rock, originally one of the successive floors of the old lake, can be followed without bend or break from chasm to chasm, and pinnacle to pinnacle. Tumultuous as the surface may be, it has no relation to underground disturbances, for the rocks are as level and unbroken as when they were laid down. It owes its ruggedness entirely to erosion.

But there is a further feature which crowns the repulsiveness of the Bad Lands. There are no springs or streams. Into the soil, parched by the fierce heats of a torrid summer, the moisture of the subsoil ascends by capillary attraction, carrying with it the saline solutions it has extracted from the rocks. At the surface it is at once evaporated, leaving behind a white crust or efflorescence, which covers the bare ground and encrusts the pebbles strewn thereon. Vegetation wholly fails, save here and there a bunch of salt-weed or a bush of the ubiquitous sage-brush, the parched livid green of which serves only to increase the desolation of the desert.

How, then, has this strange type of landscape been produced? The rainfall is exceedingly small, though from time to time come heavy showers that no doubt do much to furrow the crumbling sides of the cliffs and "buttes," and sweep down the detritus to lower ground. The main instrument of destruction, however, is not rain. In the clear dry air of these western regions the daily range of temperature is astonishingly great. In my own experience the thermometer rose sometimes to 90° in the shade, and fell at night to 19° Fahr. But this daily range of 71° is much exceeded. Exposed during the day to the expansion caused by such heat, and during the night to contraction from such rapid chilling, the surface of the friable strata is in a constant state of strain, under which it exfoliates and crumbles into sand. The sultry air during the earlier part of the day remains motionless. Again and again we saw mirage across the plains. The isolated buttes and projecting cliffs were broken up into clumps like trees, beneath which lay what seemed the sheen of a placid lake, though really a parched sage-brush plain, or a burning expanse of sand and alkali soil. But in the afternoon a wind always rose and swept across the country, though fortunately, during our exploration, never getting beyond a breeze. But it was not difficult to realise what these blasts must be in the full blaze of summer, when the hot air, like the breath of a simoom, rushes along the desert, lifting up clouds of sand and of the fine white efflorescent dust. The powdery surface of the crumbling rocks is blown away. Wastes of loose sand, here piled into shifting dunes, there dispersed far and wide over the desert, are continually augmented by fresh supplies of material from the same source. Every pebble that projects above the ground acquires, under the action of the ceaseless sand-drift, a curiously polished and channelled surface. And the same erosive action no doubt affects the mouldering precipices of the Bad Lands. The rocks are actually ground down by their own detritus, driven against them by the wind.

To the south of the Bad Lands lie the Uintah Mountains, one of the most interesting ranges in North America; for, instead of following the usual north and south direction, it runs nearly east and west, and, in place of a central crystalline wedge driven through the younger formations, it consists of a vast flat arch of nearly horizontal strata that plunge steeply down into the plains on either side. We made an excursion from Fort Bridger into these mountains. From the arid plains the change was pleasant to the densely forest-clad flanks of the chain. We had, as guide, from the Judge, an old trapper who had long hunted in the mountains, and who had a good wallet of stories for the camp-fire at night. We shall not soon forget our first day's experience of an American forest. Starting early with the view of getting above the timber-line, and having a general bird's-eye view of the interior of the mountains, we rode for several hours through the forest, making for a far peak that rose high above the dense forest of pine. Probably the first remark of a novice from the Old World, when he enters the forests of the New, is suggested by the slimness and height of the trees; they look like huge poles, feathered at top, and stuck irregularly into the ground—sometimes so near each other that one cannot force one's way between two trunks. Rarely, even in the opener glades, does one meet with a really handsome, well-grown stem, throwing its branches out freely all the way up. The next subject of astonishment is the variety of stages of growth among the timber. The tiny sapling, not long enough for a walking-stick, may be seen springing up beside the mouldering prostrate stem of a departed patriarch of the forest. Between these extremes every gradation may be seen at any place where one chooses to look, giving an impression of calm undisturbed nature and venerable antiquity. Another novelty, and perhaps the most striking of all, is the sight of so much fallen timber. Many trees die and decay, but yet remain erect, either because their roots hold, or because their stems are kept in place by the support of their still living neighbours. Others lose their stability, and topple over upon those next them. Every angle of inclination among these decaying stems may be observed. You can ride below some of them, though with the risk of having your hat switched off by some unobserved branch. Others you may walk your horse over, and an animal accustomed to the work acquires wonderful dexterity in surmounting these obstacles. But when the trunks approach the ground, or when they lie piled across each other, as they so continually do, you must ride round them; so that in those parts of the forest where fallen timber is plentiful your progress becomes provokingly slow and laborious. To us, however, everything was fresh. We rode on, hour after hour, in a kind of new world, gradually ascending till we found ourselves on the crest of a wide valley filled with pine-forest up to the brim, yet with stripes of green meadow peeping out here and there along its centre. From the farther side of this great depression rose the fine snow-streaked summits of the chain. The descent was less easy than the ascent had been, for the trees had fallen thickly down the steep declivity, which was further roughened by rocky ledges and fallen crags that would have been easy enough to surmount with free hands and feet, but which acquired in our eyes a novel importance from the difficulty of getting a horse over them. Nevertheless, every obstacle was successfully overcome. We climbed the opposite side of the valley, as far as it was practicable to take the horses, and then, leaving them in charge of "Dan," scaled the crags and steep slopes of *débris*. We were soon above the limit of tree-growth, and emerged at last on a broad bare plateau between 11,000 and 12,000 feet above the sea.

The structure of the Uintah Mountains has been investigated by several surveying parties under the Engineer and Interior Departments. Having read the reports of the Hayden, Powell, and King surveys, I was now able to take in, with comparative ease, the general aspect and meaning of the magnificent panorama around us. The broad central mass of the range is constructed of a flat arch of dull-red sandstones. The isolated peaks and ranges of buttressed cliffs along this part of the mountains reveal everywhere the horizontality of their component strata. Like the Bad Lands, but on a far more magnificent scale, they have been cut into their present forms by atmospheric sculpturing. Originally the rocks stretched in an unbroken sheet across the mountains; but in the course of ages this continuous mantle has been enormously eroded. Deep and wide valleys, vast amphitheatres, lofty terraced alcoves, and profound gorges, fretted with an infinite array of peaks, buttresses, pinnacles, columns, obelisks, and endless forms which defy the observer to find properly descriptive names for them, have gradually been carved out of these rocks. Isolated cones, with singularly majestic architectural forms, have been left standing in the midst of the denudation as monuments of its greatness. The world can show few more impressive memorials of the efficacy of subaërial erosion than the Uintah Mountains. There are no structureless crystalline rocks here to deceive us with their ruggedness. Every peak and crest, valley and cañon, bears witness to superficial sculpture. Wherever the eye turns it detects the same long lines of horizontal stratification that serve as a base from which the reality and amount of the erosion may be measured. To gain such a vivid impression of the importance of subaërial waste in the evolution of mountain-forms was worth all the long journey in itself. Yet to the south of these mountains, in the high plateaux of Utah and the great basin of the Colorado, the proofs of enormous superficial waste rise to such a gigantic scale as wholly to baffle every observer who has yet attempted to describe them.

A little below the summit which we had gained we found some bushes in fruit that recalled the wild gooseberry of home; near these a few stunted Douglas pines struggled for life. But of animal life at these heights we neither saw nor heard any sign, though bears, deer, and other large game haunt the surrounding forests. Rejoining the horses and then descending as rapidly as possible, we passed on the way some little tarns filling high recesses of the mountain, but so thickly wooded round that we failed to find the ice-worn sides that were no doubt there to mark the presence of a former glacier; for no sooner had we reached the valley-bottom than abundant traces of vanished glaciers made their appearance in the form of perfect crescent-shaped moraine mounds thrown across the valley. On these were strewn huge blocks of red sandstone, borne of old on the surface of the ice from far crags on the sky-line. Each mound of rubbish had served as a more or less effective barrier in the pathway of the stream, ponding back its waters into a lake that had eventually been converted into a meadow. But far more effective than the glacier-made dams had been those of the beaver. The extent to which the valley bottoms in this and the other mountain ranges of Western North America have been changed by the operations of this animal is almost incredible. In a single valley, for example, hundreds of acres are gradually submerged, and their cotton-wood or other tree-growth is killed. In this way the floor of the valley is cleared of timber. The beaver-ponds eventually silting up, become first marshes and then by degrees fine meadows. Riding along the stream we passed on its banks several groups of short stakes thrust into the ground and tied together so as to form a framework as if for low huts or wigwams. They were quite deserted, and had been so for some time. Dan told us they were constructed by the Indians for bathing purposes. Each of them is large enough to hold only one person at a time. When in use they are covered with skins, a fire is kindled inside and kept burning until a few stones placed in it are thoroughly warmed. The Indian or his squaw then creeps in, remains until perspiration has been induced, and finally dashes out into the stream below. It was curious to find this simple form of the sudatorium and frigidarium among the Utes in the wilds of the Far West.

It was now afternoon. We rested near an old beaver-dam, caught a few trout for supper, and crossing the valley began the ascent of its farther side. The point at which we recrossed the stream was considerably lower than that by which we had made our way in the morning. But I had taken my bearings when we were clear of the timber and had no doubt we should strike into our previous route. The ascent was steeper, rougher, and more impeded with fallen timber than anything we had yet come to. By the time we reached the summit the golden sunlight was playing in level beams among the tall pines of the crest, and we knew it would be dark in little more than an hour. Pushing on through the forest, our guide kept more and more towards the right hand, away from the line which I felt sure was that of my bearings from the mountain. We should have reached our camp, or at least the valley leading to it, but there was no sign of either. Nothing all round us but a forest that was growing every minute darker and more hopeless. At last Dan, who would not admit that he had lost his way, consented, but with some show of reluctance, to wheel round to the left. Night was now descending fast. Here and there we emerged from the gloom of the pines into an open space where there had been a forest fire. Seen in the dim light of departing day, tall trunks blackened by the fire, others bleached white by the loss of their scorched barks, rose up like a company of spectres, swinging their gaunt arms against the sky as if to warn us not to pass them into the darkness beyond. After such opener intervals the forest, as we re-entered it, became more sombre than ever. The trees seemed to close all around and over us. The fallen timber increased in confusion, the horses stumbled on, and we could no longer see to guide them. Reaching at last a little glade above which we could see the stars, we resolved to pass the night there. Dan took charge of the horses, and we groped our way to where we hoped to find water. Our search proved successful, and as we were tired and thirsty we drank heartily from some pools which we could not see, and only discovered by getting into them. On our return we found that Dan had kindled a fire, which was blazing and crackling merrily. This was nearly all the comfort that could be had unde

the circumstances. For we had no food with us except
the trout caught in the afternoon, and no covering for the
night save the saddle-cloths of the horses. There was no
help for it, however; so the trout were duly roasted and
eaten, and each donned his saddle-cloth as bed and bedding
combined. But before long it was evident that, choosing
his fireplace in the dark, our guide had placed it in rather
perilous proximity to a quantity of dried brushwood and
fallen timber. And, indeed, before we could do anything
to prevent them, the flames spread onward till a venerable
pine caught fire, and was soon a sheet of coruscating fire-
works. His neighbours followed his example, and in a few
minutes it was evident that the forest was on fire. The
flames rushed along the branches, mounting higher and
higher far up into the lofty crests of the pines, whence
showers of sparks flew out and fell in long lines through
the profoundly calm air. Tree after tree joined the
conflagration, till the reports of the exploding branches, the
hiss of the leaping flames, and the crash of the falling fire-
brands, with the ghastly glare that now died down almost
to darkness and anon shot forth into renewed brightness,
made sleep unwelcome even had it been willing to come.
Fortunately the fire eventually spent its fury on the trees
that stood round the open spot we had selected. It had
died down before morning. The presence of so much heat
around us did little to modify the cold of the night air,
and our thin saddle-cloths were not of much more service.
My friend and I huddled as close together as possible,
and lay looking up at the quiet stars as they slowly sailed
across our little space of sky, yet keeping an eye, too, on
the progress of the conflagration, lest by any chance the
flames should spread and surround us. The stones underneath us seemed somehow to grow harder and more promi-

nent before morning. I got up more than once and removed an offending block, but its place was soon taken by another. At last the first faint blush of dawn appeared beyond the pine-tops. As soon as daylight returned the horses, which had been labouring wearily all night to find a meal among the brushwood, were harnessed, and we resumed the march. It was a glorious morning. Not a breath of air was yet astir. Long wreaths of blue smoke from our conflagration lay at rest among the pine-trees, like streaks of cloud asleep on a mountain. We followed the same line that we had been pursuing when darkness came down the evening before. We had gone scarcely half a mile when we found ourselves at the edge of an open valley, and there in front stood our tent, gleaming white in the morning sunlight.

X.

THE GEYSERS OF THE YELLOWSTONE.[1]

THE traveller by railway across the American continent, after traversing several hundred miles of barren plain and sandy desert, finds at last that the line begins sensibly to descend. The panting engine moves along with increasing ease and diminished noise as its enters a long valley that leads out of the western plains, sweeping by the base of high cliffs, past the mouths of narrow lateral valleys, crossing and recrossing the watercourses by slim creaking bridges; now in a deep cutting, now in a short tunnel, it brings picturesque glimpses into view in such quick succession as almost to weary the eye that tries to scan them as they pass. After the dusty monotonous prairie, to see and hear the rush of roaring rivers, to catch sight of waterfalls leaping down the crags, scattered pine-trees crowning the heights, and green meadows carpeting the valleys, to find, too, that every mile brings you farther into a region of cultivated fields and cheerful homesteads, is a pleasure not soon to be forgotten. The Mormons have given a look of long-settled comfort to these valleys. Fields, orchards, and hedgerows, with neat farm buildings, and gardens full of flowers, remind one of bits of the old country rather than

[1] *Macmillan's Magazine*, 1881.

of the bare flowerless settlements in the West. But the sight of a group of Chinamen here and there at work on the line dispels the momentary illusion.

Winding rapidly down a succession of gorges or cañons (for every valley in the West seems to be known as a cañon), the traveller finds at last that he has entered the "Great Basin" of North America, and has arrived near the margin of the Great Salt Lake. Looking back, he perceives that the route by which he has come is one of many transverse valleys hollowed out of the flanks of the noble range of the Wahsatch Mountains. This range serves at once as the western boundary of the plateau country and as the eastern rim of the Great Basin, into which it plunges as a colossal rampart from an average height of some 4000 feet above the plain, though some of its isolated summits rise to more than twice that altitude. From the base of this great mountain-wall the country stretches westward as a vast desert plain, in a slight depression of which lies the Great Salt Lake. By industriously making use of the drainage from their mountain barrier, the Mormons have converted the strip of land between the base of the heights and the edge of the water into fertile fields and well-kept gardens.

Everybody knows that the Great Basin has no outlet to the ocean: yet nobody can see the scene with his own eyes and refuse to admit the sense of strange novelty with which it fills his mind. One's first desire is naturally to get to the lake. From a distance it looks blue enough, and not different from other sheets of water. But on a nearer view its shore is seen to be a level plain of salt-crusted mud. So gently does this plain slip under the water that the actual margin of the lake is not very sharply drawn. The water has a heavy, motionless, lifeless aspect, and is practically destitute of living creatures of every kind.

Fish are found in the rivers leading into the lake, but into the lake itself they never venture. Nor did we see any of the abundant bird-life that would have been visible on a fresh-water lake of such dimensions. There was a stillness in the air and on the water befitting the strange desert aspect of the scenery.

After looking at the water for a while, the next step was of course to get into it. The Mormons and Gentiles of Salt Lake City make good use of their lake for bathing purposes. At convenient points they have thrown out wooden piers provided with dressing-rooms and hot-water apparatus. Betaking ourselves to one of these erections, my companion and I were soon fitted out in bathing costumes of approved pattern, and descending into the lake at once realised the heaviness of the water. In walking, the leg that is lifted off the bottom seems somehow bent on rising to the surface, and some exertion is needed to force it down again to the mud below. One suddenly feels top-heavy, and seems to need special care not to turn feet uppermost. The extreme shallowness of the lake is also soon noticed. We found ourselves at first barely over the knees; so we proceeded to march into the lake. After a long journey, so long that it seemed we ought to be almost out of sight of the shore, we were scarcely up to the waist. At its deepest part the lake is not more than about fifty feet in depth. Yet it measures eighty miles in length by about thirty-two miles in extreme breadth. We made some experiments in flotation, but always with the uncomfortable feeling that our bodies were not properly ballasted for such water, and that we might roll over or turn round head downmost at any moment. It is quite possible to float in a sitting posture with the hands brought round the knees. As one of the risks of these experiments, moreover, the water would now and then get into our eyes, or find out any half-healed wound which the blazing sun of the previous weeks had inflicted upon our faces. So rapid is the evaporation in the dry air of this region that the skin after being wetted is almost immediately crusted with salt. I noticed, too, that the wooden steps leading up to the pier were hung with slender stalactites of salt from the drip of the bathers. After being pickled in this fashion we had the luxury of washing the salt crust off with the *douche* of hot water wherewith every dressing-room is provided.

It was strange to reflect that the varied beauty of the valleys in the neighbouring mountains, with their meadows, clumps of cottonwood trees, and rushing streams, should lead into this lifeless stagnant sea. One could not contemplate the scene without a strong interest in the history of the Great Salt Lake. The details of this history have been admirably worked out by Mr. G. K. Gilbert. Theoretically, we infer that the salt lakes of continental basins were at first fresh, and have become salt by the secular evaporation of their waters and consequent concentration of the salt washed into them from their various drainage basins. But in the case of the Great Salt Lake the successive stages of this long process have been actually traced in the records left behind on the surface of the ground. At present the amount of water poured into the lake nearly balances the amount lost by evaporation, so that, on the whole, the level of the lake is maintained. There are, however, oscillations of level dependent, no doubt, upon variations of rainfall. When the lake was surveyed by the Fortieth Parallel Survey in 1872, its surface was found to be eleven feet higher than it was in 1866. During the last few years, on the other hand, the lake has been diminishing. The Mormons have had to build additions to the ends of their bathing piers, from which the water had receded. There has been considerable anxiety, too, at Salt Lake City, on the subject of the diminished rainfall, which has seriously affected the supply of water for irrigation and other purposes.

That the aspect of this part at least of the Great Basin was formerly widely different is conclusively proved by some singular features, which are among the first to attract the notice even of the non-scientific traveller as he journeys round the borders of the lake. Along the flanks of the surrounding mountains there runs a group of parallel level lines, so level indeed that when first seen they suggest some extensive system of carefully-engineered waterways. On a far larger scale they are the equivalents of our well-known Parallel Roads of Glen Roy. Mile after mile they can be followed, winding in and out along the mountain declivities, here and there expanding where a streamlet has pushed out a cone of detritus, and again narrowing to hardly perceptible selvages along steeper rocky faces, but always keeping their horizontality and their proper distance from each other. That these terraces are former shore-lines of the lake admits of no doubt. The highest of them is 940 feet above the present surface of the lake, which is 4250 feet above the sea. Hence, when the lake stood at the line of that terrace, its surface was 5190 feet above sea-level. Now, it has been found that the highest terrace corresponds with a gap in the rim of the basin lying considerably to the north of the existing margin of the lake. Consequently, when the lake stood at its highest level it had an outlet northwards into the Snake River, draining into the Pacific Ocean, and must thus have been fresh. Moreover, search in the deposits of the highest terrace has brought to light convincing proof of the freshness of the water at that time for numerous shells have been found belonging to lacus trine species. At its greatest development the lake must have been vastly larger than now—a huge inland sea of fresh water lying on the western side of the continent, and quite comparable with some of the great lakes on the eastern side. It measured about 300 miles from north to south, and 180 miles in extreme width from east to west. Into this great reservoir of fresh water fishes from the tributary rivers no doubt freely entered, so that, on the whole, a community of species would be established

Fig. 25.—Terraces of Great Salt Lake, along the flanks of the Wahsatch Mountains, south of Salt Lake City.

throughout the basin. But when, owing to diminution of the rainfall, the lake no longer possessed an outlet, and in the course of ages grew gradually salt, it became unfit for the support of life. Ever since this degree of salinity was reached the rivers have been cut off from any communication with each other. These are precisely the conditions which the naturalist most desires in tracing the progress of change in animal forms. During a period which, in a geological sense, is comparatively short, but which, measured by years, must be of long duration, each river-basin has been an isolated area, with its own peculiarities of rock-structure, slope, vegetation, character of water, food, and other conditions of environment that tell so powerfully on the evolution of organic types. A beginning has been made in working out the natural history of these basins; but much patient labour will be needed before the story can be adequately told. There are probably few areas in the world which offer to the student of evolution so promising a field of research.

In the course of my brief sojourn in the region, I was able to make an observation of some interest in regard to the history of the former wide enlargement of the Great Salt Lake. The Wahsatch Mountains, which rise so picturesquely above the narrow belt of Mormon cultivation between their base and the edge of the water, have their higher parts more or less covered, or at least streaked, with snow even in midsummer, though at the time of my visit, by reason of the great heat, and I suppose in part also of a diminished snowfall, the snow had almost entirely disappeared. But any cause which could lower the mean summer temperature a few degrees would keep a permanent snow-cap on the summits, and a little further decrease would send glaciers down the valleys. That glaciers formerly did descend from the central masses of the Wahsatch range is put beyond question by the scored and polished rocks, and the huge piles of moraine detritus which they have left behind them. These phenomena have been described by the geologists of the Fortieth Parallel Survey, and I could fully confirm their observations. But I further noticed at the Little Cottonwood Cañon that the moraines descend to the edge of the highest terrace, and that the glacial rubbish forms part of the alluvial deposits there. Hence we may infer that at the time of the greatest extension of the lake the Wahsatch Mountains were a range of snowy alps, from which glaciers descended to the edge of the water. Salt Lake City being nearly on the same parallel of latitude with Naples, the change to the former topography would be somewhat as if a lofty glacier-bearing range took the place of the Apennines in the South of Europe.

One leading object of our journey was to see the wonders of the Yellowstone—that region of geysers, mud volcanoes, hot springs and sinter-beds, which the United States Congress, with wise forethought, has set apart from settlement and reserved for the instruction of the people. In a few years this part of the continent will no doubt be readily accessible by rail and coach. At the time of our visit it was still difficult of approach. We heard on the way the most ominous tales of Indian atrocities committed only a year or two before, and were warned to be prepared for something of the kind in our turn. So it was with a little misgiving as to the prudence of the undertaking that we struck off from the line of the Union Pacific Railway at Ogden and turned our faces to the north. Ogden is the centre at which the railway from Salt Lake City and that from Northern Utah and Idaho join the main trans-continental line. The first part of the journey passed pleasantly enough. The track is a very narrow one, and the carriages are proportionately small. We started in the evening, and sitting at the end of the last car enjoyed the glories of a sunset over the Great Salt Lake. Next day about noon brought us to the end of the railway in the midst of a desert of black basalt and loose sand, with a tornado blowing the hot desert dust in blinding clouds through the air. It was the oddest "terminus" conceivable, consisting of about a score of wooden booths stuck down at random, with rows of freight waggons mixed up among them, and a miscellaneous population of a thoroughly Western kind. In a fortnight afterwards the railway was to be opened some fifty miles farther north, and the whole town and its inhabitants would then move to the new terminus. Some weeks afterwards, indeed, we returned by rail over the same track, and the only traces of our mushroom town were the tin biscuit-boxes, preserved-meat cans, and other *débris* scattered about on the desert and too heavy for the wind to disperse.

With this cessation of the railway all comfort in travelling utterly disappeared. A "stage," loaded inside and outside with packages, but supposed to be capable of carrying eight passengers besides, was now to be our mode of conveyance over the bare, burning, treeless, and roadless desert. The recollection of those two days and nights stands out as a kind of nightmare. I gladly omit further reference to them. There should have been a third day and night, but by what proved a fortunate accident we escaped this prolongation of the horror. Reaching Virginia City (!), a collection of miserable wooden houses, many of them deserted — for the gold of the valley is exhausted, though many Chinese are there working over the old refuse heaps — we learnt that we were too late for the stage to Boseman. Meeting, however, a resident from Boseman as anxious to be there as ourselves, we secured a carriage and were soon again in motion. By one of the rapid meteorological changes not infrequent at such altitudes, the weather, which had before been warm, and sometimes even hot, now became for a day or two disagreeably chilly. As we crossed a ridge into the valley of the Madison River, snow fell, and the mountain crests had had their first whitening for the season when we caught sight of them, peak beyond peak, far up into the southern horizon. This valley contained the first illustrations we had yet seen of those vast alluvial accumulations which formed so marked a feature of many of the larger rivers of Western America where they debouch from the mountains. Across the whole broad plain, evidently of alluvial origin, the Madison had worked its way from side to side. From the mouths of the principal tributary valleys higher terraces of alluvium opened out upon the main valley, each affluent projecting a tongue of detritus from the base of the hills (Fig. 26). Night had fallen when we crossed the Madison River below its last

Fig. 26.— Alluvial Cones of the Madison Valley.

cañon, and further progress became impossible. There was a "ranch," or cattle-farm, not far off, where our companion had slept before, and where he proposed that we should demand quarters for the night. A good-natured welcome reconciled us to rough fare and hard beds.

On the afternoon of the third day we at length reached Boseman, the last collection of houses between us and the Yellowstone. A few miles beyond it stands Fort Ellis, a post of the United States army, built to command an important pass from the territory to the east still haunted by Indians. Through the kind thoughtfulness of my friend

Dr. Hayden, I had been provided with letters of introduction from the authorities at Washington to the commandants of posts in the West. I found my arrival expected at Fort Ellis, and the quartermaster happened himself to have come down to Boseman. Before the end of the afternoon we were once more in comfort under his friendly roof. And here I am reminded of an incident at Boseman which brought out one of the characteristics of travel in America, and particularly in the West. It may be supposed that after so long and so dusty a journey our boots were not without the need of being blacked. Having had luncheon at the hotel, I inquired of the waiter where I should go to get this done. He directed me to the clerk in the office. This formidable personage, seated at his ledger, quietly remarked to me, without raising his eyes off his pen, that he guessed I could find the materials in the corner. And there, true enough, were blacking-pot and brush, with which every guest might essay to polish his boots or not, as he pleased. In journeying westward we had sometimes seen a placard stuck up in the bedrooms of the hotels to the effect that ladies and gentlemen putting their boots outside their doors must be understood to do so at their own risk. In the larger hotels a shoe-black is one of the recognised functionaries, with his room and chair of state for those who think it needful to employ him.

Of Fort Ellis and the officers' mess there, we shall ever keep the pleasantest memories. No Indians had now to be kept in order. There was indeed nothing to do at the Fort save the daily routine of military duty. A very small incident in such circumstances is enough to furnish amusement and conversation for an evening. We made an excursion into the hills to the south, and had the satisfaction of starting a black bear from a cover of thick herbage almost below our feet. Not one of the party happened to have brought a rifle, and the animal was rapidly out of reach of our revolvers, as he raced up the steep side of the valley, and took refuge among the crags and caves of limestone at the top.

Being assured that the Yellowstone country was perfectly safe, that we should probably see no Indians at all, and that any who might cross our path belonged to friendly tribes, and being further anxious to avoid having to return and repeat that dismal stage journey, we arranged to travel through the "Yellowstone Park," as it is termed, and through the mountains encircling the head-waters of the Snake River, so as to strike the railway not far from where we had left it. This involved a ride of somewhere about 300 miles through a mountainous region still in its aboriginal loneliness. By the care of Lieutenant Alison, the quartermaster of the Fort, and the liberality of the army authorities, we were furnished with horses and a pack-train of mules, under an escort of two men, one of whom, Jack Bean by name, had for many years lived among the wilds through which we were to pass, as trapper and miner by turns; the other, a soldier in the cavalry detachment at the Fort, went by the name of "Andy," and acted as cook and leader of the mules. The smaller the party, the quicker could we get through the mountains, and as rapidity of movement was necessary, we gladly availed ourselves of the quartermaster's arrangements. Provisions were taken in quantity sufficient for the expedition, but it was expected we should be able to add to our larder an occasional haunch of antelope or elk, which in good time we did. So, full of expectation, we bade adieu, not without regret, to our friends at Fort Ellis, and set out upon our quest.

The reader may be reminded here that the Yellowstone

River has its head-waters close to the watershed of the continent, among the mountains which, branching out in different directions, include the ranges of the Wind River, Owl Creek, Shoshonee, the Tetons, and other groups that have hardly yet received names. Its course at first is nearly north, passing out of the lake where its upper tributaries collect their drainage, through a series of remarkable cañons till about the latitude of Fort Ellis, after which it bends round to the eastward, and eventually falls into the Missouri. We struck the river just above its lowest cañon in Montana. It is there already a noble stream, winding through a broad alluvial valley, flanked with hills on either side, those on the right or east bank towering up into one of the noblest ranges of the Rocky Mountains. Here, as well as on the Madison, we met with illustrations on a magnificent scale of the general law of valley structure, that every gorge formed by the convergence of the hills on either side has an expansion of the valley into a lake-like plain on its upper side. For several hours we rode along this plain among mounds of detritus, grouped in that crescent-shaped arrangement so characteristic of glacier-moraines. Large blocks of crystalline rock, quite unlike the volcanic masses along which we were travelling, lay tossed about among the mounds. One mass in particular, lying far off in the middle of the valley, looked at first like a solitary cottage. Crossing to it, however, we found it to be only a huge erratic of the usual granitoid gneiss. There could be no doubt about the massiveness of the glaciers that once filled up the valley of the Yellowstone. The moraine mounds extend across the plain and mount the bases of the hills on either side. The glacier which shed them must consequently have been here a mile or more in breadth. All the way up the valley we were on the outlook for evidence as to the thickness of the ice, which might be revealed by the height at which either transported blocks had been stranded or a polished and striated surface had been left upon the rocks of the valley. We were fortunate in meeting with evidence of both kinds.

I shall not soon forget my astonishment on entering the second cañon. We had made our first camp someway farther down, and before striking the tent in the morning had mounted the hills on the left side and observed how

Fig. 27.—Terraces below the second cañon of the Yellowstone.

the detritus (glacial detritus, as we believed it to be) had been re-arranged and spread out into terraces (Fig. 27), either by the river when at a much higher level than that at which it now flows, or by a lake which evidently once filled up the broad expansion of the valley between the two lowest cañons. We were prepared, therefore, for the discovery of still more striking proof of the power and magnitude of the old glaciers, but never anticipated that so gigantic and perfect a piece of ice-work as the second cañon was in store for us. From a narrow gorge, the sides of which rise to heights of 1000 feet or more, the river darts out into the plain which we had been traversing. The rocky sides of this ravine are smoothly polished and striated from the bottom up apparently to the top. Some of the detached knobs of schist rising out of the plain at the mouth of the cañon were as fresh in their ice-polish as if the glacier had only recently retired from them. The scene reminded me more of the valley of the Aar above the Grimsel than of any other European glacier-ground. As we rode up the gorge with here and there just room to pass between the rushing river and the rocky declivity, we could trace the ice-worn bosses of schist far up the heights till they lost themselves among the pines. The frosts of winter are slowly effacing the surfaces sculptured by the vanished glacier. Huge angular blocks are from time to time detached from the crags and join the piles of detritus at the bottom. But where the ice-polished surfaces are not much traversed with joints they have a marvellous power of endurance. Hence they may have utterly disappeared from one part of a rock-face and remain perfectly preserved on another adjoining part. There could be no doubt now that the Yellowstone glacier was massive enough to fill up the second cañon to the brim, that is to say, it must have been there at least 800 or 1000 feet thick. But in the course of our ascent we obtained proof that the thickness was even greater than this, for we found that the ice had perched blocks of granite and gneiss on the sides of the volcanic hills not less than 1600 feet above the present plain of the river, and that it not merely filled up the main valley, but actually over-rode the bounding hills so as to pass into some of the adjacent valleys. That glaciers once nestled in these mountains might have been readily anticipated, but it was important to be able to demonstrate their former existence, and to show that they attained such a magnitude.

The glaciers, however, were after all an unexpected or incidental kind of game. We were really on the trail of volcanic productions, and devoted most of our time to the hunt after them. The valley of the Yellowstone is of high antiquity. It has been excavated partly out of ancient crystalline rocks, partly out of later stratified formations, and partly out of masses of lava that have been erupted during a long succession of ages. Here and there it has been invaded by streams of basalt, which have subsequently been laboriously cut through by the river. In the whole course of our journey through the volcanic region we found that the oldest lavas were trachytes and their allies, while the youngest were as invariably basalts, the interval between the eruption of the two kinds having sometimes been long enough to permit the older rocks to be excavated into gorges before the emission of the more recent. Even the youngest, however, must have been poured out a long while ago, for they too have been deeply trenched by the slow erosive power of running water. But the volcanic fires are not yet wholly extinguished in the region. No lava, indeed, is now emitted, but there are plentiful proofs of the great heat that still exists but a short way below the surface.

Quitting the moraine mounds of the Yellowstone Valley, which above the second cañon become still more abundant and perfect, we ascended the tributary known as Gardiner's River, and camped in view of the hot springs. The first glimpse of this singular scene, caught from the crest of a dividing ridge, recalls the termination of a glacier. A mass of snowy whiteness protrudes from a lateral pine-clad valley, and presents a steep front to the narrow plain at its base. The contrast between it and the sombre hue of the pines all round heightens the resemblance of its form and aspect to a mass of ice. It is all solid rock, however, deposited by the hot water which, issuing from innumerable openings down the valley, has in course of time filled it up with this white sinter. Columns of steam rising from the mass bore witness, even at a distance, to the nature of the locality. We wandered over this singular accumulation, each of us searching for a pool cool enough to be used as a bath. I found one where the water, after quitting its conduit, made a circuit round a basin of sinter, and in so doing cooled down sufficiently to let one sit in it. The top of the mound, and indeed those parts of the deposit generally from which the water has retreated, and which are therefore dry and exposed to the weather, are apt to crack into thin shells or to crumble into white powder. But along the steep front, from which most of the springs escape, the water collects into basins at many different levels. Each of these basins has the most exquisitely fretted rim. It is at their margins that evaporation proceeds most vigorously and deposition takes place most rapidly, hence the rim is being constantly added to. The colours of these wavy, frill-like borders are sometimes remarkably vivid. The sinter, where moist or fresh, has a delicate pink or salmon-coloured hue that deepens along the edge of each basin into rich yellows, browns, and reds. Where the water has trickled over the steep front from basin to basin, the sinter has assumed smooth curved forms like the sweep of unbroken waterfalls. At many points, indeed, as one scrambles along that front, the idea of a series of frozen waterfalls rises in the mind. There are no eruptive springs or geysers at this locality now, though a large pillar of sinter on the plain below probably marks the site of one. Jack assured us that even since the time he had first been up here, some ten years before, the water had perceptibly diminished.

The contrast between the heat below and the cold above ground at nights was sometimes very great. We used to rise about daybreak, and repairing to the nearest brook or river for ablution, sometimes found a crust of ice on its quiet pools. One night, indeed, the thermometer fell to 19°, and my sponge, lying in its bag inside our tent, was solidly frozen, so that I could have broken it with my hammer. The camping-ground, selected where wood, water, and forage for the animals could be had together, was usually reached by about three o'clock in the afternoon, so that we had still several hours of daylight for sketching, or any exploration which the locality seemed to invite. About sunset Andy's fire had cooked our dinner, which we set out on the wooden box that held our cooking implements. Then came the camp-fire stories, of which our companions had a sufficient supply. Andy, in particular, would never be outdone. Nothing marvellous was told that he could not instantly cap with something more wonderful still that had happened in his own experience. What distances he had ridden! What hairbreadth escapes from Indians he had gone through! What marvels of nature he had seen! And all the while, as the tales went round and the fire burnt low or was wakened into fiercer blaze by piles of pine logs hewn down by Jack's diligent axe, the stars were coming out in the sky overhead. Such a canopy to sleep under! Wrapping myself round in my travelling cloak, I used to lie apart for a while gazing up at that sky so clear, so sparkling, so utterly and almost incredibly different from the bleared cloudy expanse we must usually be content with at home. Every familiar constellation had a brilliancy we never see through our moisture-laden atmosphere. It seemed to swim overhead, while behind and beyond it the heavens were aglow with stars that are hardly ever visible here at all. These quiet half-hours with the quiet stars, amid the silence of the primeval forest, are among the most delightful recollections of the journey.

Our mules were a constant source of amusement to us and of execration to Jack and Andy. Andy led the party with his loaded rifle slung in front of his saddle ready for any service. After him came the string of mules with their packs, followed by Jack, who with volleys of abuse and frequent applications of a leathern saddle-trap, endeavoured to keep up their pace and preserve them in line. My friend and I varied our position, sometimes riding on ahead and having the pleasure of first starting any game that might be in our way, more frequently lingering behind to enjoy quietly some of the delicious glades in the forest. But we could never get far out of hearing of the whack of Jack's belt or the fierce whoop with which he would ever and anon charge the rearmost mules and send them scampering on till every spoon, knife, and tin-can in the boxes rattled and jingled. The proper packing of a mule is an art that requires considerable skill and practice, and Jack was a thorough master of the craft. After breakfast he used to collect the animals, while Andy made up the packs, and the two together proceeded to the packing. Such tugging and pulling and kicking on the part of men and mules! The quadrupeds, however, whatever their feelings might be, 'gave no audible vent to them. But the men found relief in such fusillades of swearing as I had never before heard or even imagined. I ventured one morning to ask whether the oaths were a help to them in the packing. Jack assured me that if I had them mules to pack he'd give me two days, and at the end of that he'd bet I'd swear myself worse than any of them. Another morning Andy was hanging his coat on a branch projecting near the camp fire. The coat, however, fell off the branch and was, as a matter of course, greeted by its owner with an execration. It was put up again, and again slipped down. This was repeated two or three times, and each time the language was getting fiercer and louder. At last, when the operation was successfully completed, I asked him of what use all the swearing at the coat had been. "Wall, boss," rejoined he triumphantly, "don't ye see the darned thing's stuck up now?" This I felt was, under the circumstances, an unanswerable argument. Western teamsters are renowned for their powers of continuous execration. I myself heard one swear uninterruptedly for about ten minutes at a man who was not present, but who, it seemed, was doomed to the most horrible destruction, body and soul, as soon as this bloodthirsty ruffian caught sight of him again, either in this world or the next.

From Gardiner's River we made a *détour* over a long ridge dotted with ice-borne blocks of granite and gneiss, and crossed the shoulder of Mount Washburne by a col 8867 feet above the sea, descending once more to the Yellowstone River at the head of the Grand Cañon. The whole of this region consists of volcanic rocks, chiefly trachytes, rhyolites, obsidians, and tuffs. We chose as our camping-ground a knoll under a clump of tall pines, with a streamlet of fresh water flowing below it in haste to join the main river, which, though out of sight, was audible in the hoarse thunder of its falls. Impatient to see this ravine, of whose marvels we had heard much, we left the mules rolling on the ground and our packers getting the camp into shape, and struck through the forest in the direction of the roar. Unprepared for anything so vast, we emerged from the last fringe of the woods, and stood on the brink of the great chasm silent with amazement.

The Grand Cañon of the Yellowstone is a ravine from 1000 to 1500 feet deep. Where its shelving sides meet at the bottom there is little more than room for the river to flow between them, but it widens irregularly upwards. It has been excavated out of a series of volcanic rocks by the flow of the river itself. The waterfalls, of which there are here two, have crept backward, gradually eating their way out of the lavas and leaving below them the ravine of the Grand Cañon. The weather has acted on the sides of the gorge, scarping some parts into precipitous crags, and scooping others back, so that each side presents a series of projecting bastions and semi-circular sloping recesses. The dark forests of pine that fill the valley above sweep down to the very brink of the gorge on both sides. Such is the general plan of the place; but it is hardly possible to convey in words a picture of the impressive grandeur of the scene.

We spent a long day sketching and wandering by the side of the cañon. Scrambling to the edge of one of the bastions and looking down, we could see the river far below dwarfed to a mere silver thread. From this abyss the crags and slopes towered up in endless variety of form, and with the weirdest mingling of colours. Much of the rock, especially of the more crumbling slopes, was of a pale sulphur yellow. Through this groundwork harder masses of dull scarlet, merging into purple and crimson, rose into craggy knobs and pinnacles, or shot up in sheer vertical walls. In the sunlight of the morning the place is a blaze of strange colour, such as one can hardly see anywhere save in the crater of an active volcano. But as the day wanes the shades of evening, sinking gently into the depths, blend their livid tints into a strange mysterious gloom, through which one can still see the white gleam of the rushing river and hear the distant murmur of its flow. Now is the time to see the full majesty of the cañon. Perched on an outstanding crag one can look down the ravine and mark headland behind headland mounting out of the gathering shadows and catching up on their scarred fronts of yellow and red the mellower tints of the sinking sun. And above all lie the dark folds of pine sweeping along the crests of the precipices, which they crown with a rim of sombre green. There are gorges of far more imposing magnitude in the Colorado Basin, but for dimensions large enough to be profoundly striking, yet not too vast to be taken in by the eye at once, for infinite changes of picturesque detail, and for brilliancy and endless variety of colouring, there are probably few scenes in the world more impressive than the Grand Cañon of the Yellowstone. Such at least were the feelings with which we reluctantly left it to resume our journey.

The next goal for which we made was the Geyser Basin of the Firehole River—a ride of two days, chiefly through forest, but partly over bare volcanic hills. Some portions of this ride led into open parklike glades in the forest, where it seemed as if no human foot had ever preceded us; not a trail of any kind was to be seen. Here and there, however, we noticed footprints of bears, and some of the trees had their bark plentifully scratched at a height of three or four feet from the ground, where, as Jack said, "the bears had been sharpening their claws." Deer of different kinds were not uncommon, and we shot enough to supply our diminishing larder. Now and then we came upon a skunk or a badger, and at night we could hear the mingled bark and howl of the wolves. Andy's rifle was always ready, and he blazed away at everything. As he rode at the head of the party the first intimation those behind had of any game afoot was the crack of his rifle, followed by the immediate stampede of the mules and a round of execration from Jack. I do not remember that he ever shot anything save one wild duck, which immediately sank, or at least could not be found.

Reaching at length the Upper Geyser Basin, we camped by the river in the only group of trees in the immediate neighbourhood that had not been invaded by the sheets of white sinter which spread out all round on both sides of the river. There were hot springs, and spouting geysers, and steaming caldrons of boiling water in every direction. We had passed many openings by the way whence steam issued. In fact, in some parts of the route we seemed to be riding over a mere crust between the air above and a huge boiling vat below. At one place the hind foot of one of the horses went through this crust, and a day or two afterwards, repassing the spot, we saw it steaming. But we had come upon no actual eruptive geyser. In this basin, however, there is one geyser which, ever since the discovery of the region some ten years ago, has been remarkably regular in its action. It has an eruption once every hour and a few minutes more. The kindly name of "Old Faithful" has accordingly been bestowed upon it. We at once betook ourselves to this vent. It stands upon a low mound of sinter, which, seen from a little distance, looks as if built up of successive sheets piled one upon another. The stratified appearance, however, is due to the same tendency to form basins so marked at the Hot Springs on Gardiner's River. These basins are bordered with the same banded, brightly-coloured rims which, running in level lines, give the stratified look to the mound. On the top the sinter has gathered into huge dome-shaped or coral-like lumps, in the midst of which lies the vent of the geyser—a hole not more than a couple of feet or so in diameter—

Fig. 28.—"Old Faithful" in eruption.

whence steam constantly issues. When we arrived a considerable agitation was perceptible. The water was surging up and down a short distance below, and when we could not see it for the cloud of vapour its gurgling noise remained distinctly audible. We had not long to wait before the water began to be jerked out in occasional spurts. Then suddenly, with a tremendous roar, a column of mingled water and steam rushed up for 120 feet into the air, falling in a torrent over the mound, the surface of which now streamed with water, while its strange volcanic colours glowed vividly in the sunlight. A copious stream of still steaming water rushed off by the nearest channels to the river. The whole eruption did not last longer than about five minutes, after which the water sank in the funnel, and the same restless gurgitation was resumed. Again, at the usual interval, another outburst of the same kind and intensity took place.

Though the most frequent and regular in its movements, "Old Faithful" is by no means the most imposing of the geysers either in the volume of its discharge or in the height to which it erupts. The "Giant" and "Beehive" both surpass it, but are fitful in their action, intervals of several days occurring between successive explosions. Both of them remained tantalisingly quiet, nor could they be provoked by throwing stones down their throats to do anything for our amusement. The "Castle Geyser," however, was more accommodating. It presented us with a magnificent eruption. A far larger body of water than at "Old Faithful" was hurled into the air, and continued to rise for more than double the time. It was interesting to watch the rocket-like projectiles of water and steam that shot through and out of the main column, and burst into a shower of drops outside. At intervals, as the energy of discharge oscillated, the column would sink a little, and then would mount up again as high as before, with a 'hiss and roar that must have been audible all round the geyser basin, while the ground near the geyser perceptibly trembled. I had been sketching close to the spot when the eruption began, and in three minutes the place where I had been sitting was the bed of a rapid torrent of hot water rushing over the sinter floor to the river.

Without wearying the reader with details that possess interest only for geologists, I may be allowed to refer to one part of the structure of these geyser mounds which is not a little curious and puzzling—the want of sympathy between closely adjacent vents. At the summit of a mound the top of the subterranean column of boiling water can be seen about a yard from the surface in a constant state of commotion, while at the base of the mound, at a level thirty or forty feet lower, lie quiet pools of steaming water, some of them with a point of ebullition in their centre. There can be no direct connection between these pipes. Their independence is still more strikingly displayed at the time of eruption, for while the geyser is spouting high into the air these surrounding pools go on quietly boiling as before. It is now generally acknowledged that the seat of eruptive energy is in the underground pipe itself, each geyser having its peculiarities of shape, depth, and temperature. But it would appear also that at least above this seat of activity there can be no communication even between contiguous vents on the same geyser mound.

Another interesting feature of the locality is the tendency of each geyser to build up a cylinder of sinter round its vent. A few of these are quite perfect, but in most cases they are more or less broken down, as if they had been blown out by occasional explosions of exceptional severity. Usually there is only one cylindrical excrescence on a sinter mound; but in some cases several may be seen with their bases almost touching each other. As the force of the geyser diminishes and its eruptions become less frequent the funnel seems to get choked up with sinter, until in the end the hollow cylinder becomes a more or less solid pillar. Numerous eminences of this kind are to be seen throughout the region. Their surfaces are white and crumbling. They look, in fact, so like pillars of salt that one could not help thinking of Lot's wife, and wondering whether such geyser columns could ever have existed on the plains of Sodom. In a rainless climate they might last a long time. But the sinter here, as at Gardiner's River, when no longer growing by fresh deposits from the escaping water, breaks up into thin plates. Those parts of the basin where this disintegration is in progress look as if they had been strewn with pounded oyster-shells.

That the position of the vents slowly changes is indicated on the one hand by the way in which trees are spreading from the surrounding forest over the crumbling floor of sinter, and on the other by the number of dead or dying trunks which here and there rise out of the sinter. The volcanic energy is undoubtedly dying out. Yet it remains still vigorous enough to impress the mind with a sense of the potency of subterranean heat. From the upper end of the basin the eye ranges round a wide area of bare sinter plains and mounds, with dozens of columns of steam rising on all sides; while even from among the woods beyond an occasional puff of white vapour reveals the presence of active vents in the neighbouring valley. A prodigious mass of sinter has, in the course of ages, been laid down, and the form of the ground has been thereby materially changed. We made some short excursions into the forest, and as far as we penetrated the same floor of sinter was everywhere traceable. Here and there a long extinct geyser mound was nearly concealed under a covering of vegetation, so that it resembled a gigantic ant-hill; or a few steaming holes about its sides or summit would bring before us some of the latest stages in geyser history.

One of the most singular sights of this interesting region are the mud volcanoes, or mud geysers. We visited one of the best of them, to which Jack gave the name of "the Devil's Paint-pot." It lies near the margin of the Lower Geyser Basin. We approached it from below, surmounting by the way a series of sinter mounds dotted with numerous vents filled with boiling water. It may be described as a huge vat of boiling and variously-coloured mud, about thirty yards in diameter. At one side the ebullition was violent, and the grayish-white mud danced up into spurts that were jerked a foot or two into the air. At the other side, however, the movement was much less vigorous. The mud there rose slowly into blister-like expansions, a foot or more in diameter, which gradually swelled up till they burst, and a little of the mud with some steam was tossed up, after which the bubble sank down and disappeared. But nearer the edge on this pasty side of the caldron the mud appeared to become more viscous, as well as more brightly-coloured green and red, so that the blisters when formed remained, and were even enlarged by expansion from within, and the ejection of more liquid mud over their sides. Each of these little cones was in fact a miniature volcano with its circular crater atop. Many of them were not more than a foot high. Had it been possible to transport one unbroken, we could easily have removed it entire from its platform of hardened mud. It would have been something to boast of, that we had brought home a volcano. But, besides our invincible abhorrence of the vandalism that would in any way disturb these natural productions, in our light marching order the specimen, even had we been barbarous enough to remove it, would soon have been reduced to the condition to which the jolting of the mules had brought our biscuits—that of fine powder. We remained for hours watching the formation of these little volcanoes, and thinking of Leopold von

Buch and the old exploded "crater of elevation" theory. Each of these cones was nevertheless undoubtedly a true crater of elevation.

Willingly would we have lingered longer in this weird district. But there still lay a long journey before us ere we again could reach the confines of civilisation; we had therefore to resume the march. The Firehole River, which flows through the Geyser Basins, and whose banks are in many places vaporous heaps of sinter, the very water of the river steaming as it flows along, is the infant Madison River, which we had crossed early in the journey, far down below its lowest cañon, on our way to Fort Ellis. Our route now lay through its upper cañon, a densely-timbered gorge with picturesque volcanic peaks mounting up here and there on either side far above the pines. Below this defile the valley opens out into a little basin, filled with forest to the brim, and then, as usual, contracts again towards the opening of the next cañon. We forded the river, and, mounting the ridges on its left side, looked over many square miles of undulating pine-tops,—a vast dark-green sea of foliage stretching almost up to the summits of the far mountains. At last, ascending a short narrow valley full of beaver dams, we reached a low flat watershed 7063 feet above the sea, and stood on the "great divide" of the continent. The streams by which we had hitherto been wandering all ultimately find their way into the Missouri and the Gulf of Mexico; but the brooks we now encountered were some of the infant tributaries of the Snake or Columbia River, which drains into the Pacific. Making our way across to Henry's Fork, one of the feeders of the Snake River, we descended its course for a time. It led us now through open moor-like spaces, and then into seemingly impenetrable forest. For some time the sky towards the west had been growing more hazy as we approached, and we now found out the cause. The forest was on fire in several places. At one part of the journey we had just room to pass between the blazing crackling trunks and the edge of the river. For easier passage we forded the stream, and proceeded down its left bank, but found that here and there the fire had crossed even to that side. Most of these forest fires result from the grossest carelessness. Jack was particularly cautious each morning to see that every ember of our camp fire was extinguished, and that by no chance could the dry grass around be kindled, for it might smoulder on and slowly spread for days, until it eventually set the nearest timber in a blaze. We used to soak the ground with water before resuming our march. These forest fires were of course an indication that human beings, either red or white, had been on the ground not long before us. But we did not come on their trail. One morning, however—it was the last day of this long march—we had been about a couple of hours in the saddle. The usual halt had been made to tighten the packs, and we were picking our way across a dreary plain of sage-brush on the edge of the great basalt flood of Idaho, when Jack, whose eyes were like a hawk's for quickness, detected a cloud of dust far to the south on the horizon. We halted, and in a few minutes Jack informed us that it was a party of horsemen, and that they must be Indians from their way of riding. As they came nearer we made out that there were four mounted Indians with four led horses. Jack dismounted and got his rifle ready. Andy, without saying a word, did the same. They covered with their pieces the foremost rider, who now spurred on rapidly in front of the rest, gesticulating to us with a rod or whip he carried in his hand. "They are friendly," remarked Jack, and down went the rifles. The first rider came up to us, and after a palaver with Jack, in which we caught here and there a word of broken English, we learnt that they were bound for a council of Indians up in Montana.

Four more picturesque savages could not have been desired to complete our reminiscences of the Far West. Every bright colour was to be found somewhere in their costumes. One wore a bright blue coat faced with scarlet; another had chosen his cloth of the tawniest orange. Their straw hats were encircled with a band of down and surmounted with feathers. Scarlet braid embroidered with beads wound in and out all over their dress. Their rifles (for every one of them was fully armed) were cased in richly-broidered canvas covers, and were slung across the front of their saddles, ready for any emergency. One of them, the son of a chief, whose father Jack had known, carried a twopenny looking-glass hanging at his saddle-bow. We were glad to have seen the noble savage in his war-paint among his native wilds. Our satisfaction, however, would have been less had we known then what we only discovered when we got down into Utah, that a neighbouring tribe of the Utes were in revolt, that they had murdered the agent and his people, and killed a United States officer and a number of his soldiers, who had been sent to suppress the rising, and that there were rumours of the disaffection spreading into other tribes. We saluted our strangers with the Indian greeting, "How!" whereupon they gravely rode round and formally shook hands with each of us. Jack, however, had no faith in Indians, and after they had left us, and were scampering along the prairie in a bee-line due north, he still kept his eye on them till they entered a valley among the mountains, and were lost to sight. In half an hour afterwards another much larger cloud of dust crossed the mouth of a narrow valley down which we were moving. Waiting a little unperceived, to give the party time to widen their distance from us, we were soon once more upon the great basalt plain.

The last section of our ride proved to be in a geological sense one of the most interesting parts of the whole journey. We found that the older trachytic lavas of the hills had been deeply trenched by lateral valleys, and that all these

Fig. 29.—View on the Snake River. Basalt Plain with younger volcanic cones.

valleys had a floor of the black basalt that had been poured out as the last of the molten materials from the now extinct volcanoes. There were no visible cones or vents from which these floods of basalt could have proceeded. We rode for hours by the margin of a vast plain of basalt, stretching southward and westward as far as the eye could reach. It seemed as if the plain had been once a great lake or sea of molten rock which surged along the base of the hills, entering every valley, and leaving there a solid floor of bare black stone. We camped on this basalt plain near some springs of clear cold water which rise close to its edge. Wandering over the bare hummocks of rock, on many of which not a vestige of vegetation had yet taken root, I realised with vividness the truth of an assertion made first by Richthofen, but very generally neglected by geologists, that our modern volcanoes, such as Vesuvius or Etna, present us with by no means the grandest type of volcanic action, but rather belong to a time of failing activity. There have been periods of tremendous volcanic energy, when, instead of escaping from a local vent, like a Vesuvian cone, the lava has found its way to the surface by innumerable fissures opened for it in the solid crust of the globe over thousands of square miles. I felt that the structure of this and the other volcanic plains of the Far West furnish the true key to the history of the basaltic plateaux of Ireland and Scotland, which had been an enigma to me for many years.

At last we reached the railway that had been opened only a week or two before. Andy rode on ahead to the terminus, to intimate that we wished to be picked up. In a short while the train came up, and as we sat there in the bare, desolate valley, the engine slowed at sight of us. Our two companions were now to turn back and take a shorter route to Fort Ellis, but would be at least ten days on the march. We parted from them not without regret. Rough, but kindly, they had done everything to make the journey a memorably pleasant one to us. We took our seats in the car, and from the window, as we moved away, caught the last glimpse of our cavalcade, Andy in front with a riderless horse, and Jack in the rear with another.

XI.

THE LAVA-FIELDS OF NORTH-WESTERN EUROPE.[1]

FROM the earliest times of human tradition the basin of the Mediterranean has been the region from which our ideas of volcanoes and volcanic action have been derived. When the old classical mythology passed away and men began to form a more intelligent conception of a nether region of fire, it was from the burning mountains of that basin that the facts were derived which infant philosophy sought to explain. Pindar sang of the crimson floods of fire that rolled down from the summit of Etna to the sea as the buried Typhœus struggled under his mountain load. Strabo, with matter-of-fact precision and praiseworthy accuracy, described the eruptions of Sicily and the Æolian Islands, and pointed out that Vesuvius, though it had never been known as an active volcano, yet bore unequivocal marks of having once been corroded by fires that had eventually died out from want of fuel. In later centuries, as the circle of human knowledge and experience widened, it has still been by the Mediterranean type that the volcanic phenomena of other countries have been judged. When a geologist thinks or writes of volcanoes

[1] *Nature*, November 1880.

and volcanic action, it is the structure and products of such mountains as Etna and Vesuvius that are present to his mind. Nowhere over the whole surface of the globe have eruptions been witnessed different in kind from those of the Mediterranean vents, though varying greatly in degree. And hence even among those who have specially devoted themselves to the study of volcanoes there has been a tacit assumption that from the earliest times and in all countries of the world where volcanic outbreaks have occurred, it has been from local vents like those of Etna, the Æolian Islands, the Phlegræan Fields, or the Greek Archipelago.

If one were to assert that this assumption is probably erroneous, that the type of volcanic "cones and craters" has not been in every geological age and all over the earth's surface the prevalent one; that, on the contrary, it is the less portentous, though possibly always the more frequent type of volcanic action, and belongs perhaps to a feebler or waning degree of volcanic excitement—these statements would be received by most European geologists with incredulity, if not with some more pronounced form of dissent. Yet I am convinced that they are well founded, and that a striking illustration of their truth is supplied by the greatest of all the episodes in the volcanic history of Europe, that of the basalt-plateaux of the north-west.

It is now some twelve years since Richthofen pointed out that on the Pacific slope of North America there is evidence of the emission of vast floods of lava without the formation of cones and craters. Geologists interested in these matters may remember with what destructive energy Scrope reviewed that writer's *Natural System of Volcanic Rocks;* how he likened it to the old crude notions that had been in vogue in his own younger days, and which a study of the classical district of Auvergne had done so much to dispel; how he ridiculed what he regarded as "fanciful ideas" and "untenable distinctions," which it was "a miserable thing" to find still taught in mining schools abroad. My own reverence for the teaching of so eminent a master and so warm-hearted a friend led me to acquiesce without question in the dictum of the author of *Considerations on Volcanoes*. Having rambled over Auvergne with his admirable sections and descriptions in my hand, I knew his contention as to the removal of cones and craters by denudation and the survival of more or less fragmentary plateaux once connected with true cones to be undoubtedly correct with respect at least to that region. Nevertheless there were features of former volcanic action on which the phenomena of modern volcanoes seemed to me to throw very little light. In particular, the vast number of fissures which in Britain had been filled with basalt and now formed the well-known and abundant "dykes," appeared hardly to connect themselves with any known phase of volcanism. The area over which these dykes can be traced is probably not less than 100,000 square miles, for they occur from Yorkshire to Orkney, and from Donegal to the mouth of the Tay. As they pierce formations of every age, including the Chalk, as they traverse even the largest faults and cross from one group of rocks into another without interruption or deflection, as they become more numerous towards the great basaltic plateaux of Antrim and the Inner Hebrides, and as they penetrate the older portions of these plateaux, I inferred that the dykes probably belonged to the great volcanic period which witnessed the outburst of these western basalts. Further research has fully confirmed this inference. There can be no doubt that the outpouring of these great floods of lava of which the hills of Antrim, Mull, Morven, Skye, Faröe, and part of Iceland are merely surviving fragments and the extravasation of these thousands of dykes are connected manifestations of volcanic energy during the Tertiary period.

But this association of thin nearly level sheets of basalt piled over each other to a depth of sometimes 3000 feet, with lava-filled fissures sometimes 200 miles distant from them, presented difficulties which in the light of modern volcanic action remained insoluble. The wonderfully persistent course and horizontality of the basalts with the absence or paucity of interstratified tuffs, and the want of any satisfactory evidence of the thickening and uprise of the basalts towards what might be supposed to be the vents of eruption, were problems which again and again I attempted vainly to solve. Nor so long as the incubus of "cones and craters" lies upon one's mind does the question admit of an answer. A recent journey in Western America has at last lifted the mist from my geological vision. Having travelled for many leagues over some of the lava-fields of the Pacific slope, I have been enabled to realise the conditions of volcanism described by Richthofen, and, without acquiescing in all his theoretical conclusions, to judge of the reality of the distinction which he rightly drew between "massive eruptions" and ordinary volcanoes with cones and craters. Never shall I forget an afternoon in the autumn of last year upon the great Snake River lava desert of Idaho. It was the last day of a journey of several hundred miles through the volcanic region of the Yellowstone and Madison. We had been riding for two days over fields of basalt, level as lake-bottoms among the valleys, and on the morning of the last day, after an interview with an armed party of Indians (it was only a few days before the disastrous expedition of Major Thornburgh, and, unknown to us, the surrounding tribes were already in a ferment), we emerged from the mountains upon the great sea of black lava which seems to stretch illimitably westwards. With minds keenly excited by the incidents of the journey, we rode for hours by the side of that apparently boundless plain. Here and there a trachytic spur projected from the hills, succeeded now and then by a valley up which the black flood of lava would stretch away into the high grounds. It was as if the great plain had been filled with molten rock which had kept its level and wound in and out along the bays and promontories of the mountain-slopes as a sheet of water would have done. Copious springs and streams which issue from the mountains are soon lost under the arid basalt. The Snake River itself, however, has cut out a deep gorge through the basalt down into the trachytic lavas underneath, but winds through the desert without watering it. The precipitous walls of the cañon show that the plain is covered by a succession of parallel sheets of basalt to a depth of several hundred feet. Here and there, I was told, streams that have crossed from the hills and flowed underneath the lava desert issue at the base of the cañon walls, and swell the Snake River on its way to the Pacific. The resemblance of the horizontal basalt sheets of this region to those with which I was familiar at home brought again vividly before my mind the old problem of our Miocene dykes and Richthofen's rejected type of "massive" or fissure eruptions. I looked round in vain for any central cone from which this great sea of basalt could have flowed. It assuredly had not come from the adjacent mountains, which consisted of older and very different lavas, round the worn flanks of which the basalt had eddied. A few solitary cinder cones rose at wide intervals from the basalt plain, as piles of scoriæ sometimes do from the vapour vents on the surface of a Vesuvian lava-stream, and were as unequivocally of secondary origin (Fig. 29). Riding hour after hour among these arid wastes, I became convinced that all volcanic phenomena are not to be explained by the ordinary conception of volcanoes, but that there is another and grander type of volcanic action, where, instead of issuing from a local vent, whether or not along a line of fissure, and piling up a cone of lava and ashes around it, the molten rock has risen in many fissures, accompanied by the discharge of little or no fragmentary material, and has welled forth so as to flood the lower ground with successive horizontal sheets of basalt. Recent renewed examination of the basalt plateaux and associated dykes in the west of Scotland has assured me that this view of their origin and connection, which first suggested itself to my mind on the lava-plains of Idaho, furnishes the true key to their history.

The date of these lava-floods of the Snake River is in a geological sense quite recent. They have been poured over the bottoms of the present valleys, sealing up beneath sheets of solid stone, river-beds and lake-floors with their layers of gravel and silt. The surface of the lava is in many places black and bare, as if it had cooled only a short time ago. Yet there has been time for the excavation of the Snake River cañon to a depth of 700 feet through the basalt floor of the plain. In so arid a climate, however, the denudation of this floor must be extremely slow. Much of the plain is a verdureless waste of loose sand and dust which has gathered into shifting dunes. Save in the gorges laid open by the main river and some of its tributaries, hardly any sections have yet been cut into the volcanic floor. Dykes and other protrusions of basalt occur on the surrounding hills, but the chief fissures or vents of emission are still no doubt buried beneath the lava that escaped from them.

In North-Western Europe, however, the basalt sheets were erupted as far back as Miocene or Oligocene times. Since then, exposed to many vicissitudes of geological history—subterranean movement and changes of climate, with the whole epigene army of destructive agencies, air, rain, frost, streams, glaciers, and ice-sheets—the volcanic plateaux, trenched by valleys two or three thousand feet deep and a mile or more in breadth, and stripped bodily off many a square mile of ground over which they once spread, have been so scarped and cleft that their very roots have been laid bare. Viewed in the light of the much younger basalts of the Western Territories of North America, their history becomes at last intelligible and more than ever interesting. We are no longer under the supposed necessity of finding volcanic cones vast enough to have poured forth such widespread floods of basalt. The sources of the molten rock are to be sought in those innumerable dykes which run across Britain from sea to sea, and which in this view of their relations at once fall into their place in the volcanic history of the time.

No more stupendous series of volcanic phenomena has yet been discovered in any part of the globe. We are first presented with the fact that the crust of the earth over an area which in the British Islands alone amounted to probably not less than 100,000 square miles, but which was only part of the far more extensive region that included the Faröe Islands and Iceland, was rent by innumerable fissures in a prevalent east and west or south-east and north-west direction. These fissures, whether due to sudden shocks or slow disruption, were produced with such irresistible force as to preserve their linear character and parallelism through rocks of the most diverse nature, and even across old dislocations having a throw of many thousand feet. Yet so steadily and equably did the fissuring proceed over this enormous area, that comparatively seldom was there any vertical displacement of the sides. We rarely meet with a fissure which has been made a true fault with an upthrow and downthrow side.

The next feature is the rise of molten basalt up these thousand of fissures. The most voluminous streams of lava that have issued from any modern volcanic cone appear but as a minor manifestation of volcanic activity when compared with the filling of those countless rents over so wide a region. Mining operations in the Scottish coal-fields have shown that dykes do not always reach the surface. In all parts of the country, too, examples may be observed of breaks in the continuity of dykes. The same dyke vanishes for an interval and reappears on the same line, but is doubtless continuous underneath. What proportion of the dykes ever communicated with the surface at the time of their extravasation is a question that may perhaps never be answered. It is difficult to believe that a considerable number of them did not overflow above ground even far to the east of the main and existing outflows. But so extensive has been the subsequent denudation that all trace of such superficial emission has been removed. The general surface of the country has been lowered by subaërial waste several hundred feet at least, and the dykes now protrude as hard ribs of rock across the hills.

Traced westwards the dykes increase in abundance, till at last they reach the great basaltic plateaux. MacCulloch long ago sketched them in Skye, rising through the Jurassic rocks and merging into the overlying sheets of basalt. Similar sections occur in the other islands and in the north of Ireland. The lofty mural escarpments presented by the basalt plateaux once extended far beyond the limits to which they have now been reduced. The platform from which they have been removed shows in its abundant dykes the fissures up which the successive discharges of lava rose to the surface, where they overflowed in wide level sheets like those still so fresh and little eroded in Western North America.

That there were intervals between successive outpourings of basalt is indicated by the occasional interstratification of seams of coal and shale between the different flows. These partings contain a fragmentary record of the vegetation which grew on the neighbouring hills, and which may even have sometimes found a foothold on the crumbling surface of the basalt floor until overwhelmed by fresh floods of lava. Not a trace of marine organisms has anywhere been found among these interstratifications. There is every reason to believe that the volcanic eruptions were all subaërial. Sheet after sheet was poured forth over the wide valley between the mountains of Donegal and the Outer Hebrides on the one side, and those of the north-east of Ireland and the west of Scotland on the other, until the original surface had been buried in some places 3000 feet beneath volcanic ejections.

I believe that the most stupendous outpourings of lava in geological history have been effected not by the familiar type of conical volcano, but by these less known fissure-eruptions. Both types are of course only manifestations in different degrees of the same volcanic energy. It may be said, indeed, that both are fissure-eruptions, for the more important examples of cones and craters are no doubt placed linearly on lines of fissure. It is by no means certain that the "massive" or fissure type belongs wholly to former geological periods. In particular, one is disposed to inquire whether the great Icelandic lava-floods of 1783 —the most voluminous on record—as well as some of the recent eruptions in that island, may not have been connected rather with the opening of wide-reaching fissures than with the emissions of a single volcanic cone. The reality and importance of the grander phase of volcanism marked by fissure eruptions have been recognised by some of the able geologists who in recent years have explored the Western States and Territories of the American Union. But they have not yet received due acknowledgment on this side of the Atlantic, where the lesser type of cones and craters has been regarded as that by which all volcanic manifestations must be judged. We are fortunate in possessing in the north-west of Europe so magnificent an example of fissure-eruptions, and one which has been so dissected by denudation that its whole structure can be interpreted. The grand examples on the Pacific slope of America have yet to be worked out in detail, and will no doubt cast much fresh light on the subject, more especially upon those phenomena of which in Europe the traces have been removed by denudation. But the other continents also are not without their illustrations. The basaltic plateaux of Abyssinia and the "Deccan traps" of India probably mark the sites of some of the great fissure-eruptions which have produced the lava-fields of the Old World. In their recent admirable *résumé* of the *Geology of India*, Messrs. Medlicott and Blandford describe the persistent horizontality of the vast basalt sheets of the Deccan, the absence of any associated volcanic cones or the least trace of them in that region, and the abundance of dykes in the underlying platform of older rocks, where it emerges from beneath the volcanic plateaux. They confess the difficulty of explaining the origin of such enormous outpourings of basalt by reference to any modern volcanic phenomena. Their descriptions of these Indian Cretaceous lava-floods might, however, be almost literally applied to the Miocene plateaux of North-Western Europe and to the Pliocene or recent examples of Western North America.

XII.

THE SCOTTISH SCHOOL OF GEOLOGY.[1]

FOR the first time in the history of University Education in Scotland, we are to-day met to begin the duties of a Chair specially devoted to the cultivation of Geology and Mineralogy. Though Science is of no country nor kin, it yet bears some branches which take their hue largely from the region whence they sprang, or where they have been most sedulously nurtured. Such local colourings need not be deprecated, since they are both inevitable and useful. They serve to bring out the peculiarities of each climate, or land, or people, and it is the blending of all these colourings which finally gives the common neutral tint of science. This is in a marked degree true of Geology. Each country, where any part of the science has been more particularly studied, has furnished its local names to the general nomenclature, and its rocks have sometimes served as types from which the rocks of other regions have been classified and described. The very scenery of the country, reacting on the minds of the early observers, has sometimes influenced their observations, and has thus left an impress on the general progress of the science. As we enter to-day upon

[1] The Inaugural Lecture at the opening of the Class of Geology and Mineralogy in the University of Edinburgh, 6th November 1871.

a new phase in the cultivation of Geology here, it seems most fitting that we should look back for a little at the past development of the science in this part of the British Islands.

There was a time, still within the memory of living men, when a handful of ardent original observers here in Edinburgh carried geological speculation and research to such a height as to found a new, and, in the end, a dominant school of Geology. The history of the Natural Sciences, like that of Philosophy, has been marked by epochs of activity and intervals of quiescence. One genius, perhaps, has arisen and kindled in other minds the flame that burned so brightly in his own. A time of vigorous research has ensued, but as the personal influence that evoked it has waned, a period of feebleness or torpor has been apt to ensue, and to last until the advent of some new awakening. Such oscillations of mental energy have an importance and a significance far beyond the narrow limits of the country or city in which they may have been manifested. They form part of that long and noble record of the struggle of man with the forces of nature, and deserve the thoughtful consideration of all who have joined or who contemplate joining in that struggle. I propose on the present occasion to sketch the story of one of these periods of vigorous originality, which had its rising and its setting in this city—the story of what may be called the Scottish School of Geology. I wish to place before you, in as clear a light as I can, the work which was accomplished by the founders of that school, that you may see how greatly it has influenced, and is even now influencing, the onward march of the science. I do this in no vainglorious spirit, nor with any wish to exalt into prominence a mere question of nationality. Science knows no geographical or political limits. Nor, though we may be proud of what has been achieved for Geology in this little kingdom, can we for a moment shut our eyes to the fact that these achievements are of the past, that the measure of the early promise at the beginning of this century has been but scantily fulfilled in Scotland, and that the state of the science among us here, instead of being in advance, is rather behind the time. And thus I dwell now on the example of our predecessors solely in the hope that, realising to ourselves what that example really was, we may be stimulated to follow it. The same hills and valleys, crags and ravines, remain around us which gave these great men their inspiration, and still preach to us the lessons which they were the first to understand.

The period during which the distinctively Scottish School of Geology rose and flourished may be taken as included between the years 1780 and 1825—a brief half-century. Previous to that time Geology, in the true sense of the word, can hardly be said to have existed. Steno, indeed, more than a hundred years before, had clearly shown, from the occurrence of the remains of plants and animals embedded in the solid rocks, that the present was not the original order of things, that there had been upheavals of the sea into dry land and depressions of the land beneath the sea, by the working of forces lodged within the earth, and that the memorials of these changes were preserved for us in the rocks. Seventy years later another writer of the Italian school, Lazzaro Moro, adopting and extending the conclusions of Steno, pointed to the evidence that the surface of the earth is everywhere worn away, and is repaired by the upheaving power of earthquakes, but for which the mountains and all the dry land would at last be brought beneath the level of the waves.

But none of these desultory researches, interesting and important though they were as landmarks in the progress of science, bore immediate fruit in any broad and philosophic outline of the natural history of the globe. Men were still trammelled by the belief that the date of the creation of the world and its inhabitants could not be placed farther back than some five or six thousand years, that this limit was fixed for us in Holy Writ, and that every new fact must receive an interpretation in accordance with such limitation. They were thus often driven to distort the facts or to explain them away. If they ventured to pronounce for a natural and obvious interpretation, they laid themselves open to the charge of impiety and atheism, and might bring down the unrelenting vengeance of the Church.

Such was the state of inquiry when the Scottish Geological School came into being. The founder of that school was James Hutton, a man of a singularly original and active mind, who was born at Edinburgh in 1726, and died there in 1797. Educated for the medical profession, but possessed of a small fortune, which gave him leisure to follow his favourite pursuits, he eventually devoted himself to the study of Mineralogy. But it was not merely as rare or interesting objects, nor even as parts of a mineralogical system, that he dealt with minerals. They seemed to suggest to him constant questions as to the earlier conditions of our planet, and he thus was gradually led into the wider fields of Geology and Physical Geography. Quietly working in his study here, a favourite member of a brilliant circle of society, which included such men as Black, Cullen, Adam Smith, and Clerk of Eldin, and making frequent excursions to gather fresh data and test the truth of his deductions, he at length matured his immortal *Theory of the Earth*, and published it in 1785. Associated with Hutton, rather as a friend and enthusiastic admirer than as an independent observer, was John Playfair, Professor of Natural Philosophy in this University, by whose graceful exposition the doctrines of Hutton were most widely made known to the world. His classic *Illustrations of the Huttonian Theory* is one of the most delightful books of science in our language—clear, elegant, and vivacious—a model of scientific description and argument, which I would earnestly recommend to your notice. Sir James Hall, another of this little illustrious band, had one of the most inventive minds which have ever taken up the pursuit of science in this country. His merits have never yet been adequately realised by his countrymen, though they are better appreciated in Germany and in France. He was in fact the founder of Experimental Geology, since it was he who first brought geological speculation to the test of actual physical experiment. This he accomplished in a series of ingenious researches, whereby he corroborated some of the disputed parts of the doctrines of his master, Hutton. These were the three chief leaders of the Scottish School; but to their number, as worthy but less celebrated associates, we must not omit to add the names of Mackenzie, Webb Seymour, and Allan.

It would lead me far beyond the allotted hour of lecture to attempt any adequate summary of the work achieved by each of these early pioneers of the science. It will be enough for my present purpose to sketch what were the leading characteristics of this Scottish School, and what claim it has to be remembered, not by us only, but by all to whom Geology is the subject either of serious study or of pleasant recreation.

Born in a "land of mountain and flood," the geology

THE SCOTTISH SCHOOL OF GEOLOGY.

of the Scottish School naturally dealt in the main with the inorganic part of the science, with the elemental forces which have burst through and cracked and worn down the crust of the earth. It asked the mountains of its birthplace by what chain of events they had been upheaved, how their rocks, so gnarled and broken, had come into being, how valleys and glens had been impressed upon the surface of the land, and how the various strata through which these wind had been step by step built up. It encountered no rocks like those which had arrested the notice of the early Italian geologists, charged with fossil shells, corals, and bones of fish, such as still lived in the adjoining seas, and which at once suggested the former presence of the sea over the land. Neither did it meet with deposits showing abundant traces of ancient lakes, rivers, and land-surfaces, each marked by the presence of animal and plant remains, like those which set Steno and Moro thinking. The rocks of Scotland are, as a whole, unfossiliferous. It was therefore rather with the records of physical events, unaided by the testimony of organic remains, that the Scottish geologists had to deal. Their task was to unravel the complicated processes by which the rocky crust of the earth has been built up, and by which the present varied contour of the earth's surface has been produced—to ascertain, in short, from a study of the existing economy of the world, what has been the physical history of our planet in earlier ages. The marvellous story told by the organic remains in the earth's crust had not yet been in any way conjectured.

Hitherto, while men had been accustomed to believe that the earth was but some 6000 years old, they sought in the rocks beneath and around them evidence only of the six days' creation or of the flood of Noah. Each new cosmological system was based upon that belief, and tried in various ways to reconcile the Biblical narrative with fanciful interpretations of the facts of Nature. It was reserved for Hutton to declare, for the first time, that the rocks around us reveal no trace of the beginning of things. He, too, first clearly and persistently proclaimed the great fundamental truth of Geology, that in seeking to interpret the past history of the earth as chronicled in the rocks, we must use the present economy of nature as our guide. In our investigations, " no powers," he says, " are to be employed that are not natural to the globe, no action to be admitted of except those of which we know the principle." "Nor are we to proceed in feigning causes when those appear insufficient which occur in our experience."[1] The changes of the past must be investigated in the light of similar changes now in operation. This was a guiding principle of the Scottish School, and through their influence it has become a guiding principle of modern Geology; though, under the name of "Uniformitarianism," it has unquestionably been pushed to an unwarrantable length by some of the later followers of Hutton. The appeal to Nature in her present condition for light in geological inquiry was a watchword of the Huttonians, and in the hands of one of the most illustrious of their number, Sir Charles Lyell, has been largely influential in the establishment of Geology as a truly observational science.

There were two directions in which Hutton laboured, and in each of which he and his followers constantly travelled by the light of the present order of nature—viz. the investigation of (1) changes which have transpired beneath the surface and within the crust of the earth, and (2) changes which have been effected on the surface itself.

1. That the interior of the earth was hot, and that it

[1] Hutton's *Theory of the Earth*, i. p. 160; ii. p. 549.

was the seat of powerful forces, by which solid rocks had been rent open and wide regions of land convulsed, were familiar facts, attested by every volcano and earthquake. These phenomena had been for the most part regarded as abnormal parts of the system of nature; by many writers, indeed, as well as by the general mass of mankind, they were looked upon as Divine judgments, specially sent for the punishment and reformation of the human species. To Hutton, pondering over the great organic system of the world, a deeper meaning was necessary. He felt, as Steno and Moro had done, that the earthquake and volcano were but parts of the general mechanism of our planet. But he saw, also, that they were not the only exhibitions of the potency of subterranean agencies, that in fact they were only partial and perhaps even secondary manifestations of the influence of the great internal heat of the globe, and that the full import of that influence could not be understood unless careful study were given also to the structure of the rocky crust of the earth. Accordingly he set himself for years patiently to gather and meditate over data which would throw light upon that structure and its history. The mountains and glens, river-valleys and sea-coasts of his native country were diligently traversed by him, every journey adding something to his store of materials, and enabling him to arrive continually at wider views of the general economy of nature. At one time we find him in a Highland glen searching for proofs of a hypothesis which he was convinced must be true, and, at their eventual discovery, breaking forth into such gleeful excitement that his attendant gillies concluded he must certainly have hit upon a mine of gold. At another time we read of him boating with his friends Playfair and Hall along the wild cliffs of Berwickshire, again in search of confirmation to his views, and finding, to use the words of Playfair, "palpable evidence of one of the most extraordinary and important facts in the natural history of the earth."

As a result of his wanderings and reflection, he concluded that the great mass of the rocks which form the visible part of the crust of the earth was formed under the sea, as sand, gravel, and mud are laid down there now; that these ancient sediments were consolidated by subterranean heat, and, by paroxysms of the same force, were fractured, contorted, and upheaved into dry land. He found that portions of the rocks had even been in a fused state; that granite had been erupted through other stony masses; and that the dark trap-rocks, or "whinstones" of Scotland, were likewise of igneous origin.

When the sedimentary rocks were studied in the broad way which was followed by Hutton and his associates, many proofs appeared of ancient convulsions and re-formations of the earth's surface. It was found that among the hills the strata were often on end, while on the plains they were gently inclined; and the inference was deduced by Hutton that the former series must have been broken up by subterranean commotions before the accumulation of the latter, which was derived from its *débris*. He conjectured that the later rocks would be found actually resting upon the edges of the older. His search for, and discovery of, this relation at the Siccar Point, on the Berwickshire coast, are well described by his biographer Playfair, who accompanied him, and who, dwelling on the impression which the scene had left upon himself, adds, "The mind seemed to grow giddy by looking so far into the abyss of time; and while we listened with earnestness and admiration to the philosopher who was now unfolding to us the order and series of these wonderful events, we became sensible how much farther reason may sometimes go than imagination can venture to follow. Sir James Hall afterwards, by a series of characteristically ingenious experiments, showed how the rocks of that coast-line may have been contorted by movements in the crust of the earth under great superincumbent pressure.

Hutton contended for the former molten condition of granite and of many other crystalline masses. He maintained that the combined influence of subterranean heat and pressure upon sedimentary rocks could consolidate and mineralise them, and even convert them into crystalline masses. He may thus be regarded as the founder of the modern doctrine of metamorphism, or the gradual transformation of marine sediments into the gnarled and rugged gneiss and schist of which mountains are built up. Let me quote the eulogium passed upon this part of his work in an essay by M. Daubrée, which eleven years ago was crowned with a prize by the Academy of Sciences at Paris :—" By an idea entirely new, the illustrious Scottish philosopher showed the successive co-operation of water and the internal heat of the globe in the formation of the same rocks. It is the mark of genius to unite in one common origin phenomena very different in their nature." "Hutton explains the history of the globe with as much simplicity as grandeur. Like most men of genius, indeed, who have opened up new paths, he exaggerated the extent to which his conceptions could be applied. But it is impossible not to view with admiration the profound penetration and the strictness of induction of so clear-sighted a man at a time when exact observations had been so few, he being the first to recognise the simultaneous effect of water and heat in the formation of rocks, in imagining a system which embraces the whole physical system of the globe. He established principles which, in so far as they are fundamental, are now universally admitted."

While Hutton fortified his convictions by constant appeals to the rocks themselves, his disciple Hall tested their truth in the laboratory. It is the boast of Scotland to have led the way in the application of chemical and physical experiment to the elucidation of geological history. It was objected to Hutton's theory, that if basalt and similar rocks had ever been in a melted state, they would now have been seen in the condition of glass or slag, and not with the granular or crystalline texture which they actually possess. Hall demolished this objection by melting basalt into a glass, and then, by slow cooling, reconverting it into a granular substance more or less resembling the original rock. Hutton had maintained that under enormous pressure, such as he conceived must exist beneath the ocean, or deep within the crust of the earth, even limestone itself might be melted without losing its carbonic acid. This was ridiculed by his opponents, on whom he retorted that they "judged of the great operations of the mineral kingdom from having kindled a fire and looked into the bottom of a little crucible." Hall, however, to whom fire and crucible were congenial implements, resolved to put the question to the test of experiment, and though, out of deference to his master, he delayed his task until after the death of the latter, he did at last succeed in converting limestone, under various great pressures, into a kind of marble, and even in reducing it to complete fusion, in which state it acted powerfully on other rocks. He concluded his elaborate essay on this subject with these words: "This single result affords, I conceive, a strong presumption in favour of the solution which Dr. Hutton has advanced of all the geological phenomena; for the truth of the most doubtful principle which he has assumed has thus been established by direct experiment."

Though they saw clearly the proofs which the rocks afford us of former revolutions, neither Hutton nor his friends had any conception of the existence of the great series of fossiliferous formations which has since been unfolded by the labours of later observers—that voluminous record in which the history of life upon this planet has been preserved. They spoke of "Alpine schistus," "primary" or "secondary" strata, as if the geological past had consisted but of two great ages—the second replete with traces of the destruction of the first. "The ruins of an older world," said Hutton, "are visible in the present structure of our planet." He knew nothing of the long, but then undiscovered, succession of such "ruins," each marking a wide interval of time. Nevertheless for the establishment of the great truths which Hutton laboured to confirm, such knowledge was not necessary. On the other hand, it was most needful that the significance of that discordance between the older and newer strata which Hutton recognised should be persistently proclaimed. And the Huttonians, in spite of their limited range of knowledge and opportunity, saw its value, and held by it.

2. But it was not merely, nor even perhaps chiefly, for their exposition of the structure and history of the rocks under our feet that the geologists of the Scottish School deserve to be held in lasting remembrance. They could not, indeed, have advanced as far as they did in expounding former conditions of the planet, had they not, with singular clearness, perceived the order and system of change which is in progress over the surface of the globe at the present day. It was their teaching which led men to recognise the harmony and co-operation of the forces of

nature that work within the earth, with those which are seen and felt upon its surface. Hutton first caught the meaning of that constant circulation of water which, by means of evaporation, winds, clouds, rain, snow, brooks, and rivers, is kept up between land and sea. He saw that the surface of the dry land is everywhere being wasted and worn away. The scarped cliff, the rugged glen, the lowland valley, are each undergoing this process of destruction; wherever land rises above ocean, there, from mountain-top to sea-shore, degradation is continually going on. Here and there, indeed, the *débris* of the hills may be spread out upon the plains; here and there, too, dark angular peaks and crags rise as they rose centuries ago, and seem to defy the elements. But these are only apparent and not real exceptions to the universal law, that so long as a surface of land is exposed to the atmosphere it must suffer disintegration and removal.

But Hutton saw, further, that this waste is not equally distributed over the whole face of the dry land. He perceived that while, owing to the greater or less resistance offered by different kinds of rocks, the decay must vary indefinitely in rate, its amount must necessarily be greatest where the surplus water flows off towards the sea—that is, along the channels of the streams. Watercourses, he argued, are precisely in the lines which water would naturally follow in running down the slope of the land from its water-shed to the sea, and which, when once selected by the surplus drainage, would necessarily be continually widened and deepened by the excavating power of the rivers. He regarded the streams and rivers of a country as following the lines which they had chiselled for themselves out of the solid land, and thus he arrived at the deduction that valleys have been, inch by inch and foot by foot, dug out of the solid framework of the land by the same natural agents—rain, frost, springs, rivers—by which they are still made wider and deeper. "The mountains," he said, "have been formed by the hollowing out of the valleys, and the valleys have been hollowed out by the attrition of hard materials coming from the mountains." This is a doctrine which is only now beginning to be adequately realised. Yet to Hutton it was so obvious as to convince him, to use his own memorable words, "that the great system upon the surface of this earth is that of valleys and rivers, and that however this system shall be interrupted and occasionally destroyed, it would necessarily be again formed in time while the earth continued above the level of the sea."

Although these views were again and again proclaimed by Hutton in the pages of his treatise, and though Playfair, catching up the spirit of his master, preached them with a force and eloquence which might almost have insured the triumph of any cause, they met with but scant acceptance. The men were before their time; and thus while the world gradually acknowledged the teaching of the Scottish School as to the past history of the rocks, it lent an incredulous ear to that teaching when dealing with the present surface of the earth. Even some of the Huttonians themselves refused to follow their master when he sought to explain the existing inequalities of the land by the working of the same quiet unobtrusive forces which are still plying their daily tasks around us. But no incredulity or neglect can destroy the innate vitality of truth. And so now, after the lapse of fully two generations, the views of Hutton have in recent years been revived, especially in Britain, and have become the war-cry of a yearly increasing crowd of earnest hard-working geologists.

While they insisted upon the manifest proofs of constant and universal decay over the surface of the globe, the Scottish geologists no less strongly contended that this decay was a necessary part of the present economy of nature, that it had been in progress from the earliest periods in the history of the earth, and that it was essential for the presence of organised beings upon the planet. They pointed to the vegetable soil, derived from the decomposition of the rocks which it covers, and necessary for the support of vegetable life. They appealed to the vast quantity of sedimentary rocks forming the visible part of the crust of the earth, and bearing witness in every bed and layer to the degradation and removal of former continents. They showed that the accumulated *débris* of the land, carried to sea, was there spread out on the sea-floor to form new strata, which, hardened in due time into solid rock, would hereafter be upheaved to form the frame-work of new lands.

Such was the geology of the Scottish School. It was based not on mere speculation, but on broad fundamental facts drawn from mountain and valley, hill and plain, and tested as far as was then possible by the scrutiny of actual experiment. It strove, for the first time in the history of science, to evolve a system out of the manifold complications of nature, to harmonise what had seemed but the wild random working of subterranean forces with the quiet operations in progress upon the surface of the earth, to understand what is the present system of the world, and through that to peer into the history of earlier conditions of the planet. It taught that the earthquake and volcano were parts of the orderly arrangement by which new continents were from time to time raised up to supply the place of others that had been worn away; that the surface of the land required to decay to furnish life to plants and animals; that in the removal of the *débris* thus produced mountains and valleys were carved out; and that in the depths of the ocean there were at the same time laid down the materials for the formation of other lands, which in after ages would be upheaved by underground forces, to be anew worn away as before. The Scottish School proclaimed that in the inorganic world there is ceaseless change, that this change is the central idea of the system, and that in its constant progress lie the conditions necessary for the continuance of our earth as a habitable globe.

That Hutton and his followers failed to realise that the planet has had a vastly prolonged evolution which the visible geological record chronicles only imperfectly, that they were ignorant of the geological importance of fossils, that they saw only partially the truths which they laboured so zealously to establish, and that they fell into errors, attaching to secondary and even erroneous parts of their system an importance which we now see to have been misplaced, is only what may be said of any body of men who, at any time, have led the way in a new development of human inquiry. But, after all allowance is made for such shortcomings, we see that their mistakes were, for the most part, mainly in matters of detail, and that the fundamental principles for which they fought have become the very life and soul of modern geology.

I have spoken of this Scottish School as marking a period of activity which rose into brightness and then waned. It is only too true, that so far as the originality and influence of its cultivators go, Geology has never since held in Scotland the place which it held here at the beginning of this century. Its decay is perhaps to be ascribed chiefly, if not entirely, to the introduction of the doctrines of Werner from Germany The Huttonians had dealt rather with general principles than with minute details; they were weak in accurate mineralogical knowledge—not that they were ignorant of or in any degree despised such knowledge; but it was not necessary for their object. When, however, the system of Werner came to be taught within these walls by his devoted pupil Jameson, its precision and simplicity, and its supposed capability of ready application in every country, joined to the skill and zeal of its teacher, gave it an impulse which lasted for years. I shall have occasion in a subsequent lecture to speak of this system. It attempted to explain the geological history of the globe from the rocks of a limited district in Saxony. It required mineralogical determination of rocks and pointed out a certain order of succession among them. In so far it did good service, but its theoretical teaching as regards the history of the earth cannot now be regarded without a smile. It maintained that the globe was covered with certain universal formations which had been precipitated successively from solution in a primeval ocean. Of upheaval and subsidence, earthquakes and volcanoes, and all the mechanism of internal heat, it could make nothing, and ignored as much as it dared. Werner, the founder of this system, had the faculty of attaching his students to him, and of infusing into them no small share of his own zeal and faith in his doctrines. His pupil Jameson had a similar aptitude. Skilled in the mineralogy of his time, and full of desire to apply the teachings of Freyberg to the explication of Scottish geology, or geognosy, as the Wernerians preferred to call it, Jameson gathered round him a band of active observers, who gleaned facts from all parts of Scotland, and to whom the first accurate descriptions of the mineralogy of the country are due. It is but fitting that a tribute of gratitude should on the present occasion be offered to the memory of Jameson for the lifelong devotion with which he taught Natural History, and especially Mineralogy, in this University. His influence is to be judged not by what he wrote, but by the effect of his example, and by the number of ardent naturalists who sprang from his teaching. He founded a scientific society here, and called it Wernerian, after his chief—a society which, under his guidance, did excellent service to the cause of science in Scotland. And yet in the course of my scientific reading I have never met a sadder contrast than to turn from the earlier volumes of the *Transactions of the Royal Society of Edinburgh*, containing the classic essays of Hutton, Hall, and Playfair—essays which made an epoch in the history of geology—to the pages of the *Wernerian Memoirs*, and find grave discussions about the universal formations, the aqueous origin of basalt, and the chemical disposition of such rocks as slate and conglomerate.

Between the followers of Hutton and Werner there necessarily arose a keen warfare. The one battalion of combatants was styled by its opponents "Vulcanists" or "Plutonists," as if they recognised only the power of internal fire, while the other was in turn nicknamed "Neptunists," in token of their adherence to water. The warfare lasted in a desultory way for many years, and though the Wernerian school, having essentially no vitality, eventually died out, and its leader Jameson publicly and frankly recanted its errors, the early Huttonian magnates had meanwhile one by one departed and left no successors. The Huttonian school triumphed indeed, but its triumph was seen rather in other countries than in Scotland, and was due chiefly to the impetus given to the reception of its doctrines by the *Principles of Geology* of Lyell. The

Wernerian faith preached here by Jameson attracted in great measure the younger men, and when its influence waned there were no great names on the other side to rally the thinned and weakened ranks of Huttonianism. Hence came a period of comparative quiescence, which has lasted almost down to our own day. From time to time, indeed, a geologist has arisen among us to show that the science was not dead, and that the doctrines of Hutton had borne good fruit. But geology has never since held such a prominent place in Scotland, nor have the writings of our geologists taken the same position in the literature of the science. The great name of Lyell, and others of lesser note, have earned elsewhere their title to fame.

But there is one name which must be in our hearts and on our lips to-day, that of Roderick Impey Murchison. To his munificence and the liberality of the Crown we owe the foundation of this Chair of Geology, and to his warm friendship I am indebted for the position in which I stand before you. Of his achievements in science, and of the influence of his work all over the world, it is not necessary now to speak; but on Scottish Geology no man has left his name more deeply engraven. It was he who, with Professor Sedgwick, first made known the order of succession of the Old Red Sandstone of the north of Scotland; it was he who sketched for us the relations of the great Silurian masses of the Southern uplands; and it was he who, by a series of admirable researches, brought order out of the chaos of the so-called primary rocks of the Highlands, and placed these rocks on a parallel with the Silurian strata of other countries. These labours will come again before us in detail, and you will then better understand their value, and the debt we owe to the man who accomplished them.

THE SCOTTISH SCHOOL OF GEOLOGY.

Sir Roderick Murchison looked forward with interest to the occasion which has called us together to-day. Only a few weeks ago I talked with him regarding it, and his eye brightened as I told him of the subject on which I proposed to speak to you. I had hoped that he would have lived to see this day, and to hear at least of the beginning of the work which he has inaugurated for us in this University; but this was not to be. He has been taken from us ripe in years, in work, and in honours, and he leaves us the example of his unwearied industry, his admirable powers of observation, and his rare goodness of heart.

In the course of study now before us, we are to be engaged in examining together the structure and history of the earth. We shall trace the working of the various natural agents which are now carrying on geological change, and by which the past changes of the globe may be explained. In so doing we shall be brought continually face to face with the history of life as recorded in the rocks— for it is by that history mainly that the sequences of geological time can be established. We shall thus have to trespass a little on what is the proper domain of the Professors of Botany and of Natural History. But you will find that no hard line can be drawn between the sciences. Each must needs overlap upon the other; and indeed it is in this mutual interlacing that one great element of the strength and interest of science lies. From Professors Balfour and Wyville Thomson you will learn the structure of the fossils with which we shall have to deal as our geological alphabet, and their relation to living plants and animals. By Professor Crum Brown you are taught the full meaning and application of the chemical laws under which the minerals and rocks, which we in this class must study, have been formed, and of the processes concerned in those subsequent changes, both of rocks and minerals, which are of such paramount importance in Geology.

And now, in conclusion, permit me to give expression to the feelings which must strongly possess the mind of one who is called upon to fill the first Chair dedicated in Scotland to the cultivation of Geology. When I look back to the times of that illustrious group of men—Hutton, Hall, Playfair—who made Edinburgh the special home of Geology; of Boué and MacCulloch, who gave to Scottish rocks an European celebrity; of Jameson and Edward Forbes, who did so much to stimulate the study of Geology and Mineralogy in this University; and to the memory of Hugh Miller and Charles Maclaren, who fostered the love of these sciences throughout the community, and for whose kindly friendship and guidance given to me in my boyhood I would fain express my hearty gratitude—when I cast my thoughts back upon these associations, it would be affectation to conceal the anxiety with which the prospect fills me. The memory of these great names arises continually before me, bearing with it a consciousness of the responsibility under which I lie to labour earnestly not to be unworthy of the traditions of the past. And, gentlemen, I feel deeply my responsibility to you who are to enter with me upon a yet untrodden path of the Academic curriculum. It is only experience that will show us how we shall best travel over the wide field before us. In the meantime I must bespeak your kindly forbearance. While I shall cheerfully teach you all I know, and confess what I do not know, I would fain have you in the end to regard me as much in the light of a fellow-student, searching with you after truth, as of a teacher putting before you what is already known. We have now an opportunity of combined and sedulous work which has not hitherto been obtainable in Scotland. We may not rival a Hutton or a Hall; but we may at least try to raise again the standard of geological inquiry here. On every side of us are incentives to study. Crag and hill rise around us, each eloquent of ancient revolutions, and each a silent witness of the revolution in progress now. At our very gates tower on one side the picturesque memorials of long silent volcanoes, with their crumbling lavas and ashes. On the other lies the buried vegetation of an ancient land, with the corals and shells of a former ocean. Everywhere the scarred and wasted rocks tell of the degradation of the solid land, and show us how the waste goes on. Let us then carry into our task some share of the enthusiasm which these daily exemplars called forth in bygone times. Let us turn from the lessons of the lecture-room to the lessons of the crags and ravines, appealing constantly to nature for the explanation and verification of what is taught. And thus, whatsoever may be your career in future, you will in the meantime cultivate habits of observation and communion with the free fresh world around you—habits which will give a zest to every journey, which will enable you to add to the sum of human knowledge, and which will assuredly make you wiser and better men.

XIII.

GEOGRAPHICAL EVOLUTION.[1]

In the quaint preface to his *Navigations and Voyages of the English Nation*, Hakluyt calls geography and chronology "the sunne and moone, the right eye and the left of all history." The position thus claimed for geography three hundred years ago by the great English chronicler was not accorded by his successors, and has hardly been admitted even now. The functions of the geographer and the traveller, popularly assumed to be identical, have been supposed to consist in descriptions of foreign countries, their climate, productions, and inhabitants, bristling on the one hand with dry statistics, and relieved on the other by as copious an introduction as may be of stirring adventure and personal anecdote. There has indeed been much to justify this popular assumption. It was not until the keynote of its future progress was struck by Karl Ritter, within the present century, that geography advanced beyond the domain of travellers' tales and desultory observation into that of orderly, methodical, scientific progress. This branch of inquiry, however, is now no longer the pursuit of mere numerical statistics, nor the chronicle of marvellous and often questionable adventures by flood and fell. It seeks

[1] A Lecture delivered at the Evening Meeting of the Royal Geographical Society, 24th March 1879.

to present a luminous picture of the earth's surface, its various forms of configuration, its continents, islands, and oceans, its mountains, valleys, and plains, its rivers and lakes, its climates, plants, and animals. It thus endeavours to produce a picture which shall not be one of mere topographical detail. It ever looks for a connection between scattered facts, tries to ascertain the relations which subsist between the different parts of the globe, their reactions on each other and the function of each in the general economy of the whole. Modern geography studies the distribution of vegetable and animal life over the earth's surface, with the action and reaction between it and the surrounding inorganic world. It traces how man, alike unconsciously and knowingly, has changed the face of nature, and how, on the other hand, the conditions of his geographical environment have moulded his own progress.

With these broad aims, geography comes frankly for assistance to many different branches of science. It does not, however, claim in any measure to occupy their domain. It brings to the consideration of their problems a central human interest in which these sciences are sometimes apt to be deficient; for it demands first of all to know how the problems to be solved bear upon the position and history of man and of this marvellously-ordered world wherein he finds himself undisputed lord. Geography freely borrows from meteorology, physics, chemistry, geology, zoology, and botany; but the debt is not all on one side. Save for the impetus derived from geographical research, many of these sciences would not be in their present advanced condition. They gain in vast augmentation of facts, and may cheerfully lend their aid in correlating these for geographical requirements.

In no respect does modern geography stand out more

prominently than in the increased precision and fulness of its work. It has fitted out exploratory expeditions, and in so doing has been careful to see them provided with the instruments and apparatus necessary to enable them to contribute accurate and definite results. It has guided and fostered research, and has been eager to show a generous appreciation of the labours of those by whom our knowledge of the earth has been extended. Human courage and endurance are not less enthusiastically applauded than they once were; but they must be united to no common powers of observation before they will now raise a traveller to the highest rank. When we read a volume of recent travel, while warmly appreciating the spirit of adventure, fertility of resource, presence of mind, and other moral qualities of its author, we instinctively ask ourselves, as we close its pages, what is the sum of its additions to our knowledge of the earth? From the geographical point of view—and it is to this point alone that these remarks apply—we must rank an explorer according to his success in widening our knowledge and enlarging our views regarding the aspects of nature.

The demands of modern geography are thus becoming every year more exacting. It requires more training in its explorers abroad, more knowledge on the part of its readers at home. The days are drawing to a close when one can gain undying geographical renown by struggling against man and beast, fever and hunger and drought, across some savage and previously unknown region, even though little can be shown as the outcome of the journey. All honour to the pioneers by whom this first exploratory work has been so nobly done! They will be succeeded by a race that will find its laurels more difficult to win—a race from which more will be expected, and which will need to make up in the variety, amount, and value of its detail, what it lacks in the freshness of first glimpses into new lands.

With no other science has geography become more intimately connected than with geology, and the connection is assuredly destined to become yet deeper and closer. These two branches of human knowledge are, to use Hakluyt's phrase, "the sunne and moone, the right eye and the left," of all fruitful inquiry into the character and history of the earth's surface. As it is impossible to understand the genius and temperament of a people, its laws and institutions, its manners and customs, its buildings and its industries, unless we trace back the history of that people, and mark the rise and effect of each varied influence by which its progress has been moulded in past generations; so it is clear that our knowledge of the aspect of a continent, its mountains and valleys, rivers and plains, and all its surface-features, cannot be other than singularly feeble and imperfect, unless we realise what has been the origin of these features. The land has had a history, not less than the human races that inhabit it.

One can hardly consider attentively the future progress of geography without being convinced that in the wide development yet in store for this branch of human inquiry, one of its main lines of advance must be in the direction of what may be termed geographical evolution. The geographer will no longer be content to take continents and islands, mountain chains and river valleys, tablelands and plains, as initial or aboriginal outlines of the earth's surface. He will insist on knowing what the geologist can tell him regarding the growth of these outlines. He will try to trace out the gradual evolution of a continent, and may even construct maps to show its successive stages of development. At the same time, he will seek for information regarding the history of the plants and animals of the region, and may find much to reward his inquiry as to the early migrations of the fauna and flora, including those even of man himself. Thus his pictures of the living world of to-day, as they become more detailed and accurate, will include more and more distinctly a background of bygone geographical conditions, out of which, by continuous sequence, the present conditions will be shown to have arisen.

I propose this evening to sketch in mere outline the aspects of one side of this evolutional geography. I wish to examine, in the first place, the evidence whereby we establish the fundamental fact that the present surface of any country or continent is not that which it has always borne, and the data by which we may trace backward the origin of the land; and, in the second place, to consider, by way of illustration, some of the more salient features in the gradual growth of the framework of Europe.

The first of these two divisions of the subject deals with general principles, and may be conveniently grouped into two parts: 1st, The Materials of the Land. 2d, The Building of the Land.

I. GENERAL PRINCIPLES OF CONTINENTAL EVOLUTION.

i.—*The Materials of the Land.*

Without attempting to enter into detailed treatment of this branch of the subject, we may, for the immediate purpose in view, content ourselves with the broad, useful classification of the materials of the land into two great series—Fragmental and Crystalline.

§ 1. *Fragmental.*—A very cursory examination of rocks in almost any part of the world suffices to show that by far the larger portion of them consists of compacted fragmentary materials. Shales, sandstones, and conglomerates, in infinite variety of texture and colour, are piled above each other to form the foundation of plains and the structure of mountains. Each of these rocks is composed of distinct particles, worn by air, rain, frost, springs, rivers, glaciers, or the sea, from previously existing rocks. They are thus derivative formations, and their source, as well as their mode of origin, can be determined. Their component grains are for the most part rounded, and bear evidence of having been rolled about in water. Thus we easily and rapidly reach a first and fundamental conclusion—that the substance of the main part of the solid land has been originally laid down and assorted under water.

The mere extent of the area covered by these water-formed rocks would of itself suggest that they must have been deposited in the sea. We cannot imagine rivers or lakes of magnitude sufficient to have spread over the sites of the present continents. The waters of the ocean, however, may easily be conceived to have rolled at different times over all that is now dry land. The fragmental rocks contain, indeed, within themselves proof that they were mainly of marine, and not of lacustrine or fluviatile origin. They have preserved in abundance the remains of foraminifera, corals, crinoids, molluscs, annelides, crustaceans, fishes, and other organisms of undoubtedly marine habitat, which must have lived and died in the places where their traces remain still visible.

But not only do these organisms occur scattered through sedimentary rocks; they actually themselves form thick masses of mineral matter. The Carboniferous or Mountain Limestone of Central England and Ireland, for example, reaches a thickness of from 2000 to 3000 feet, and covers thousands of square miles of surface. Yet it is almost entirely composed of congregated stems and joints and plates of crinoids, with foraminifera, corals, bryozoans, brachiopods, lamellibranchs, gasteropods, fish-teeth, and other unequivocally marine organisms. It must have been for ages the bottom of a clear sea, over which generation after generation lived and died, until their accumulated remains had gathered into a deep and compact sheet of rock. From the internal evidence of the stratified formations we thus confidently announce a second conclusion—that a great portion of the solid land consists of materials which have been laid down on the floor of the sea.

From these familiar and obvious deductions we may proceed further to inquire under what conditions these marine formations, spreading so widely over the land, were formed. According to a popular belief, shared in perhaps by not a few geologists, land and sea have been continually changing places. It is supposed that while, on the one hand, there is no part of a continent over which sea-waves may not have rolled, so, on the other hand, there is no lonely abyss of the ocean where a wide continent may not have bloomed. That this notion rests upon a mistaken interpretation of the facts may be shown from an examination—(1) of the rocks of the land, and (2) of the bottom of the present ocean.

(1) Among the thickest masses of sedimentary rock—those of the ancient palæozoic systems—no features recur more continually than alternations of different sediments, and surfaces of rock covered with well-preserved ripple-marks, trails and burrows of annelides, polygonal and irregular desiccation marks, like the cracks at the bottom of a sun-dried muddy pool. These phenomena unequivocally point to shallow and even littoral waters. They occur

from bottom to top of deposits which reach a thickness of several thousand feet. They can be interpreted only in one way, viz. that their deposition began in shallow water; that during their formation the area of deposit gradually subsided for thousands of feet; yet that the rate of accumulation of sediment kept pace on the whole with this depression; and hence, that the original shallow-water character of the deposits remained, even after the original sea-bottom had been buried under a vast mass of sedimentary matter. Now, if this explanation be true, even for the enormously thick and comparatively uniform systems of older geological periods, the relatively thin and much more varied stratified groups of later date can offer no difficulty. In short, the more attentively the stratified rocks of the crust of the earth are studied, the more striking becomes the absence of any deposits among them which can legitimately be considered those of a deep sea. They have all been deposited in comparatively shallow water.

The same conclusion may be arrived at from a consideration of the circumstances under which the deposition must have taken place. It is evident that the sedimentary rocks of all ages have been derived from degradation of land. The gravel, sand, and mud, of which they consist, existed previously as part of mountains, hills, or plains. These materials carried down to the sea would arrange themselves there as they do still, the coarser portions nearest the shore, the finer silt and mud farthest from it. From the earliest geological times the great area of deposit has been, as it still is, the marginal belt of sea-floor skirting the land. It is there that nature has always strewn "the dust of continents to be." The decay of old rocks has been unceasingly in progress on the land, and the building up of new rocks has been unintermittently going on underneath the adjoining sea. The two phenomena are the complementary sides of one process, which belongs to the terrestrial and shallow oceanic parts of the earth's surface and not to the wide and deep ocean basins.

(2) Recent explorations of the bottom of the deep sea all over the world have brought additional light to this question. No part of the results obtained by the *Challenger* Expedition has a profounder interest for geologists and geographers than the proof which they furnish that the floor of the ocean basins has no real analogy among the sedimentary formations that form most of the framework of the land. We now know by actual dredging and inspection that the ordinary sediment washed off the land sinks to the sea-bottom before it reaches the deeper abysses, and that, as a rule, only the finer particles are carried more than a few score of miles from the shore. Instead of such sandy and pebbly material as we find so largely among the sedimentary rocks of the land, wide tracts of the sea-bottom at great depths are covered with various kinds of organic ooze, composed sometimes of minute calcareous foraminifera, sometimes of siliceous radiolaria or diatoms. Over other areas vast sheets of clay extend, derived apparently from the decomposition of volcanic detritus, of which large quantities are floated away from volcanic islands, and much of which may be produced by submarine volcanoes. On the tracts farthest removed from any land the sediment seems to settle scarcely so rapidly as the dust that gathers over the floor of a deserted hall. Mr. Murray, of the *Challenger* staff, has described how from these remote depths large numbers of shark's teeth and ear-bones of whales were dredged up. We cannot suppose the number of sharks and whales to be much greater in these regions

than in others where their relics were found much less
plentifully. The explanation of the abundance of their
remains was supplied by their varied condition of decay
and preservation. Some were comparatively fresh, others
had greatly decayed, and were incrusted with or even completely buried in a deposit of earthy manganese. Yet the
same cast of the dredge brought up these different stages of
decay from the same surface of the sea-floor. While generation after generation of sea-creatures drops its bones to the
bottom, now here, now there, so exceedingly feeble is the
rate of deposit of sediment that they lie uncovered, mayhap
for centuries, so that the remains which sink to-day may lie
side by side with the mouldered and incrusted bones that
found their way to the bottom hundreds of years ago.

Another striking indication of the very slow rate at which
sedimentation takes place in these abysses has also been
brought to notice by Mr. Murray. In the clay from the
bottom he found numerous minute spherical granules of
native iron, which, as he suggests, are almost certainly of
meteoric origin — fragments of those falling stars which,
coming to us from planetary space, burst into fragments
when they rush into the denser layers of our atmosphere.
In tracts where the growth of silt upon the sea-floor is
excessively tardy, the fine particles, scattered by the dissipa-
tion of these meteorites, may remain in appreciable quantity. In this case, again, it is not needful to suppose that
meteorites have disappeared over these ocean depths more
numerously than over other parts of the earth's surface.
The iron granules have no doubt been as plentifully
showered down elsewhere, though they cannot be so readily
detected in accumulating sediment. I know no recent
observation in physical geography more calculated to impress deeply the imagination than the testimony of this

presumably meteoric iron from the most distant abysses of the ocean. To be told that mud gathers on the floor of these abysses at an extremely slow rate conveys but a vague notion of the tardiness of the process. But to learn that it gathers so slowly, that the very star-dust which falls from outer space forms an appreciable part of it, brings home to us, as hardly anything else could do, the idea of undisturbed and excessively slow accumulation.

From all this evidence we may legitimately conclude that the present land of the globe, though formed in great measure of marine formations, has never lain under the deep sea; but that its site must always have been near land. Even its thick marine limestones are the deposits of comparatively shallow water. Whether or not any trace of aboriginal land may now be discoverable, the characters of the most unequivocally marine formations bear emphatic testimony to this proximity of a terrestrial surface. The present continental ridges have probably always existed in some form, and as a corollary we may infer that the present deep ocean basins likewise date from the remotest geological antiquity.

§ 2. *Crystalline.*—While the greater part of the framework of the land has been slowly built up of sedimentary materials, it is abundantly varied by the occurrence of crystalline masses, many of which have been injected in a molten condition into rents underground, or have been poured out in lava-streams at the surface.

Without entering at all into geological detail, it will be enough for the present purpose to recognise the characters and origin of two great types of crystalline material which have been called respectively the Eruptive and Metamorphic.

(*a*) *Eruptive.*— As the name denotes, Eruptive or Igneous rocks have been ejected from the heated interior of the earth. In a modern volcano lava ascends the central funnel, and issuing from the lip of the crater or from lateral fissures pours down the slopes of the cone in sheets of melted rock. The upper surface of the lava column within the volcano is kept in constant ebullition by the rise of steam through its mass. Every now and then a vast body of steam rushes out with a terrific explosion, scattering the melted lava into impalpable dust, and filling the air with ashes and stones, which descend in showers upon the surrounding country. At the surface, therefore, igneous rocks appear, partly as masses of congealed lava, and partly as more or less consolidated sheets of dust and stones. But beneath the surface there must be a downward prolongation of the lava column, which no doubt sends out veins into rents of the subterranean rocks. We can suppose that the general aspect of the lava which consolidates at some depth will differ from that which solidifies above ground.

As a result of the revolutions which the crust of the earth has undergone, the roots of many ancient volcanoes have been laid bare. We have been, as it were, admitted into the secrets of these subterranean laboratories of nature, and have learned much regarding the mechanism of volcanic action which we could never have discovered from any modern volcano. Thus, while on the one hand we meet with beds of lava and consolidated volcanic ashes which were undoubtedly erupted at the surface of the ground in ancient periods, and were subsequently buried deep beneath sedimentary accumulations now removed, on the other hand we find masses of igneous rock which certainly never came near the surface, but must have been arrested in their ascent from below while still at a great depth, and have been laid bare to the light after the removal of the pile of rock under which they originally lay.

By noting these and other characters, geologists have learnt that, besides the regions of still active volcanoes, there are few large areas of the earth's surface where proofs of former volcanic action or of the protrusion of igneous rocks may not be found. The crust of the earth, crumpled and fissured, has been, so to speak, perforated and cemented together by molten matter driven up from below.

(*b*) *Metamorphic.*—The sedimentary rocks of the land have undergone many changes since their formation, some of which are still far from being satisfactorily accounted for. One of these changes is expressed by the term *Metamorphism*, and the rocks which have undergone this process are called *Metamorphic*. It seems to have taken place under widely varied conditions, being sometimes confined to small local tracts, at other times extending across a large portion of a continent. It consists in the rearrangement of the component materials of rocks, and notably in their recrystallisation along particular lines or laminæ. It is usually associated with evidence of great pressure; the rocks in which it occurs having been corrugated and crumpled, not only in vast folds, which extend across whole mountains, but even in such minute puckerings as can only be observed with the microscope. It shows itself more particularly among the older geological formations, or those which have been once deeply buried under more recent masses of rock, and have been exposed as the result of the removal of these overlying accumulations. The original characters of the sandstones, shales, grits, conglomerates, and limestones, of which, no doubt, these metamorphic masses once consisted, have been more or less effaced, and have given place to that peculiar crystalline laminated or foliated structure so distinctively a result of metamorphism.

An attentive examination of a metamorphic region

shows that here and there the alteration and recrystallisation have proceeded so far that the rocks graduate into granites and other so-called igneous rocks. A series of specimens may be collected showing unaltered or at least quite recognisable sedimentary rocks at the one end, and thoroughly crystalline igneous rocks at the other. Thus the remarkable fact is brought home to the mind that ordinary sandstones, shales, and other sedimentary materials may in the course of ages be converted by underground changes into crystalline granite. The framework of the land, besides being knit together by masses of igneous rock intruded from below, has been strengthened by the welding and crystallisation of its lowest rocks. It is these rocks which rise along the central crests of mountain chains, where, after the lapse of ages, they have been uncovered and laid bare, to be bleached and shattered by frost and storm.

ii.—*The Architecture of the Land.*

Let us now proceed to consider how these materials, sedimentary and crystalline, have been put together, so as to constitute the solid land of the globe.

It requires but a cursory examination to observe that the sedimentary masses have not been huddled together at random; that, on the contrary, they have been laid down in sheets one over the other. An arrangement of this kind at once betokens a chronological sequence. The rocks cannot all have been formed simultaneously. Those at the bottom must have been laid down before those at the top. A truism of this kind seems hardly to require formal statement. Yet it lies at the very foundation of any attempt to trace the geological history of a country. Did the rocks everywhere lie undisturbed one above another as they were originally laid down, their clear order of succession would carry with it its own evident interpretation. But such have been the changes that have arisen, partly from the operation of forces from below, partly from that of forces acting on the surface, that the true order of a series of rocks is not always so easily determined. By starting, however, from where the succession is normal and unbroken, the geologist can advance with confidence into regions where it has been completely interrupted; where the rocks have been shattered, crumpled, and even inverted.

The clue which guides us through these labyrinths is a very simple one. It is afforded by the remains of once living plants and animals which have been preserved in the rocky framework of the land. Each well-marked series of sedimentary accumulations contains its own characteristic plants, corals, crustaceans, shells, fishes, or other organic remains. By these it can be identified and traced from country to country across a whole continent. When, therefore, the true order of superposition of the rocks has been ascertained by observing how they lie upon each other, the succession of their fossils is at the same time fixed. In this way the sedimentary part of the earth's crust has been classified into different formations, each characterised by its distinct assemblage of organic remains. In the most recent formations, most of these remains are identical with still living species of plants and animals; but as we descend in the series and come into progressively older deposits the proportion of existing species diminishes until at last all the species of fossils are found to be extinct. Still lower and older rocks reveal types and assemblages of organisms which depart farther and farther from the existing order.

By noting the fossil contents of a formation, therefore, even in a district where the rocks have been so disturbed that their sequence is otherwise untraceable, the geologist can confidently assign their relative position to each of the fractured masses. He knows, for instance, using for our present purpose the letters of the alphabet to denote the sequence of the formations, that a mass of limestone containing fossils typical of the formation B must be younger than another mass of rock containing the fossils of A. A series of strata full of the fossils of H resting immediately on others charged with those of C, must evidently be separated from these by a great gap, elsewhere filled in by the intervening formations D, E, F, G. Nay, should the rocks in the upper part of a mountain be replete with the fossils proper to D, while those in the lower slopes showed only the fossils of E, F, and G, it could be demonstrated that the materials of the mountain had actually been turned upside down, for, as proved by its organic remains, the oldest and therefore lowest formation had come to lie at the top, and the youngest, and therefore highest, at the bottom.

Of absolute chronology in such questions science can as yet give no measure. How many millions of years each formation may have required for its production, and how far back in time may be the era of any given group of fossils, are problems to which no answer, other than a mere guess, can be returned. But this is a matter of far less moment than the relative chronology, which can usually be accurately fixed for each country, and on which all attempts to trace back the history of the land must be based.

While, then, it is true that most of the materials of the solid land have been laid down at successive periods under the sea, and that the relative dates of their deposition can be determined, it is no less certain that the formation of these materials has not proceeded uninterruptedly, and that they have not finally been raised into land by a single movement.

The mere fact that they are of marine origin shows, of course, that the land owes its origin to some kind of terrestrial disturbance. But when the sedimentary formations are examined in detail, they present a most wonderful chronicle of long-continued, oft-repeated, and exceedingly complex movements of the crust of the globe. They show that the history of every country has been long and eventful; that, in short, hardly any portion of the land has reached its present condition, save after a protracted series of geological revolutions.

One of the most obvious and not the least striking features in the architecture of the land is the frequency with which the rocks, though originally horizontal, or approximately so, have been tilted up at various angles, or even placed on end. At first it might be supposed that these disturbed positions have been assumed at random, according to the capricious operations of subterranean forces. They seem to follow no order, and to defy any attempt to reduce them to system. Yet a closer scrutiny serves to establish a real connection among them. They are found, for the most part, to belong to great, though fractured, curves, into which the crust of the earth has been folded. In low countries far removed from any great mountain range, the rocks often present scarcely a trace of disturbance, or if they have been affected, it is chiefly by having been thrown into gentle undulations. As we approach the higher grounds, however, they manifest increasing signs of commotion. Their undulations become more frequent and steeper, until, entering within the mountain region, we find the rocks curved, crumpled, fractured, inverted, tossed over each other into yawning gulf and towering crest, like billows arrested at the height of a furious storm.

Yet even in the midst of such apparent chaos it is not impossible to trace the fundamental law and order by which it is underlaid. The prime fact to be noted is the universal plication and crumpling of rocks which were at first nearly horizontal. From the gentle undulations of the strata beneath the plains to their violent contortion and inversion among the mountains, there is that insensible gradation which connects the whole of these disturbances as parts of one common process. They cannot be accounted for by any mere local movements, though such movements no doubt took place abundantly. The existence of a mountain chain is not to be explained by a special upheaval or series of upheavals caused by an expansive force acting from below. Manifestly the elevation is only one phase of a vast terrestrial movement which has extended over whole continents, and has affected plains as well as high grounds.

The only cause which, so far as our present knowledge goes, could have produced such widespread changes is a general contraction of the earth's mass. There can be no doubt that at one time our planet existed in a gaseous, then in a liquid condition. Since these early periods it has continued to lose heat, and consequently to contract and to grow more and more solid, until, as the physicists insist, it has now become practically as rigid as a globe of glass or of steel. But in the course of the contraction, after the solid external crust was formed, the inner hot nucleus has lost heat more rapidly than the crust, and has tended to shrink inward from it. As a consequence of this internal movement, the outer solid shell has sunk down upon the retreating nucleus. In so doing, it has of course had to accommodate itself to a diminished area, and this it could only accomplish by undergoing plication and fracture. Though the analogy is not a very exact one, we may liken our globe to a shrivelled apple. The skin of the apple does not contract equally. As the internal moisture passes off, and the bulk of the fruit is reduced, the once smooth exterior becomes here and there corrugated and dimpled.

Without entering into this difficult problem in physical geology, it may suffice if we carry with us the idea that our globe must once have had a greater diameter than it now possesses, and that the crumpling of its outer layers, whether due to mere contraction or, as has been suggested, to the escape also of subterranean vapours, affords evidence of this diminution. A little reflection suffices to show us that, even without any knowledge of the actual history of the contraction, we might anticipate that the effects would neither be continuous nor everywhere uniform. The solid crust would not, we may be sure, subside as fast as the mass inside. It would, for a time at least, cohere and support itself, until at last, gravitation proving too much for its strength, it would sink down. And the areas and amount of descent would be greatly regulated by the varying thickness and structure of the crust. Subsidence would not take place everywhere; for, as a consequence of the narrower space into which the crust sank, some regions would necessarily be pushed up. These conditions appear to have been fulfilled in the past history of the earth. There is evidence that the terrestrial disturbance has been renewed again and again, after long pauses, and that, while the ocean basins have on the whole been the great areas of depression, the continents have been the lines of uprise or relief, where the rocks were crumpled and pushed out of the way. Paradoxical, therefore, as the statement may appear, it is nevertheless strictly true, that the solid land, considered with reference to the earth's surface as a whole, is the consequence of subsidence rather than of upheaval.

Grasping, then, this conception of the real character of the movements to which the earth owes its present surface configuration, we are furnished with fresh light for exploring the ancient history and growth of the solid land. The great continental ridges seem to lie nearly on the site of the earliest lines of relief from the strain of contraction. They were forced up between the subsiding oceanic basins at a very early period of geological history. In each succeeding epoch of movement they were naturally used over again, and received an additional push upward. Hence we see the meaning of the evidence supplied by the sedimentary rocks as to shallow seas and proximity of land. These rocks could not have been otherwise produced. They were derived from the waste of the land, and were deposited near the land. For it must be borne in mind that every mass of land as soon as it appeared above water was at once attacked by the ceaseless erosion of moving water and atmospheric influences, and immediately began to furnish materials for the construction of future lands to be afterwards raised out of the sea.

Each great period of contraction elevated anew the much-worn land, and at the same time brought the consolidated marine sediments above water as parts of a new terrestrial surface. Again a long interval would ensue, marked perhaps by a slow subsidence both of the land and sea-bottom. Meanwhile the surface of the land was channelled and lowered, and its detritus was spread over the sea-floor, until another era of disturbance raised it once more with a portion of the surrounding ocean-bed. These successive upward and downward movements explain why the sedimentary formations do not occur as a continuous series, but often lie each upon the upturned and worn edges of its predecessors.

Returning now to the chronological sequence indicated by the organic remains preserved among the sedimentary rocks, we see how it may be possible to determine the relative order of the successive upheavals of a continent. If, for example, a group of rocks, which as before may be called A, were found to have been upturned and covered over by undisturbed beds C, the disturbance could be affirmed to have occurred at some part of the epoch represented elsewhere by the missing series B. If, again, the group C were observed to have been subsequently tilted, and to pass under gently-inclined or horizontal strata E, a second period of disturbance would be proved to have occurred between the time of C and E.

I have referred to the unceasing destruction of its surface which the land undergoes from the time when it emerges out of the sea. As a rule, our conceptions of the rate of this degradation are exceedingly vague. Yet they may easily be made more definite by a consideration of present changes on the surface of the land. Every river carries yearly to the sea an immense amount of sand and mud. But this amount is capable of measurement. It represents, of course, the extent to which the general level of the surface of the river's drainage basin is annually lowered. According to such measurements and computations as have been already made, it appears that somewhere about $\frac{1}{6000}$ of a foot is every year removed from the surface of its drainage basin by a large river. This seems a small fraction, yet by the power of mere addition it soon mounts up to a large total. Taking the mean level of Europe to be 600 feet, its surface, if everywhere worn away at what seems to be the present mean normal rate, would be entirely reduced to the sea-level in little more than three and a half millions of years.

But of course the waste is not uniform over the whole surface. It is greatest on the slopes and valleys, least on the more level grounds. A few years ago, in making some estimates of the ratios between the rates of waste on these areas, I assumed that the tracts of more rapid erosion occupy only one-ninth of the whole surface affected, and that in these the rate of destruction is nine times greater than on the more level spaces. Taking these proportions, and granting that $\frac{1}{8000}$ of a foot is the actual ascertained amount of loss from the whole surface, we learn by a simple arithmetical process that $\frac{1}{12}$ of an inch is carried away from the plains and tablelands in seventy-five years, while the same amount is worn out of the valleys in eight and a half years. One foot must be removed from the former in 10,800 years, and from the latter in 1200 years. Hence, at the present rate of erosion, a valley 1000 feet deep may be excavated in 1,200,000 years—by no means a very long period in the conception of most geologists.

I do not offer these figures as more than tentative results. They are based, however, not on mere guesses, but on data which, though they may be corrected by subsequent inquiry, are the best at present available, and are probably not far from the truth. They are of value in enabling us more vividly to realise how the prodigious waste of the land, proved by the existence of such enormous masses of sedimentary rock, went quietly on age after age, until results were achieved which seem at first scarcely possible to so slow and gentle an agency.

It is during this quiet process of decay and removal that all the distinctive minor features of the land are wrought out. When first elevated from the sea, the land doubtless presents on the whole a comparatively featureless surface. It may be likened to a block of marble raised out of the quarry — rough and rude in outline, massive in solidity and strength, but giving no indication of the grace into which it will grow under the hand of the sculptor. What art effects upon the marble block, nature accomplishes upon the surface of the land. Her tools are many and varied — air, frost, rain, springs, torrents, rivers, avalanches, glaciers, and the sea — each producing its own characteristic traces in the sculpture. With these implements, out of the huge bulk of the land she cuts the valleys and ravines, scoops the lake-basins, hews with bold hand the colossal outlines of the mountains, carves out peak and crag, crest and cliff, chisels the courses of the torrents, splinters the sides of the precipices, spreads out the alluvium of the rivers, and piles up the moraines of the glaciers. Patiently and unceasingly has this great earth-sculptor sat at her task since the land first rose above the sea, washing down into the ocean the *débris* of her labour, to form the materials for the framework of future countries; and there will she remain at work so long as mountains stand, and rain falls, and rivers flow.

II. The Growth of the European Continent.

Passing now from the general principles with which we have hitherto been dealing, we may seek an illustration of their application to the actual history of a large mass of land. For this purpose let me ask your attention to some of the more salient features in the gradual growth of Europe. This continent has not the simplicity of structure elsewhere recognisable; but without entering into detail or following a continuous sequence of events, our present purpose will be served by a few broad outlines of the condition of the European area at successive geological periods.

It is the fate of continents, no less than of the human communities that inhabit them, to have their first origin shrouded in obscurity. When the curtain of darkness begins to rise from our primeval Europe, it reveals to us a scene marvellously unlike that of the existing continent. The land then lay chiefly to the north and north-west, probably extending as far as the edge of the great submarine plateau by which the European ridge is prolonged under the Atlantic for 230 miles to the west of Ireland. Worn fragments of that land exist in Finland, Scandinavia, and the north-west of Scotland, and there are traces of what seem to have been some detached islands in Central Europe, notably in Bohemia and Bavaria. Its original height and extent can of course never be known; but some idea of them may be formed by considering the bulk of solid rock which was formed out of the waste of that land. I find that if we take merely one portion of the detritus washed from its surface and laid down in the sea —viz. that which is comprised in what is termed the Silurian system—and if we assume that it spreads over 60,000 square miles of Britain with an average thickness of 16,000 feet, or 3 miles, which is probably under the truth, then we obtain the enormous mass of 180,000 cubic miles. The magnitude of this pile of material may be better realised if we reflect that it would form a mountain ridge three times as long as the Alps, or from the North Cape to Marseilles (1800), with a breadth of more than 33 miles, and an average height of 16,000 feet—that is, higher than the summit of Mont Blanc. All this vast pile of sedimentary rock was worn from the slopes and shores of the primeval northern land. Yet it represents but a small fraction of the material so removed, for the sea of that ancient time spread over nearly the whole of Europe eastwards into Asia, and everywhere received a tribute of sand and mud from the adjoining shores.

There is perhaps no mass of rock so striking in its general aspect as that of which this northern embryo of Europe consisted. It lacks the variety of composition, structure, colour, and form, which distinguishes rocks of more modern growth; but in dignity of massive strength it stands altogether unrivalled. From the headlands of the Hebrides to the far fjords of Arctic Norway it rises up grim and defiant of the elements. Its veins of quartz, felspar, and hornblende project from every boss and crag like the twisted and knotted sinews of a magnificent torso. Well does the old gneiss of the north deserve to have been made the foundation-stone of a continent.

What was the character of the vegetation that clothed this earliest prototype of Europe is a question to which at present no definite answer is possible. We know, however, that the shallow sea which spread from the Atlantic southward and eastward over most of Europe was tenanted by an abundant and characteristic series of invertebrate animals —trilobites, graptolites, cystideans, brachiopods, and cephalopods, strangely unlike, on the whole, to anything living in our waters now, but which then migrated freely along the shores of the Arctic land between what are now America and Europe.

The floor of this shallow sea continued to sink, until over Britain, at least, it had gone down several miles. Yet the water remained shallow because the amount of sediment constantly poured into it from the north-west filled it up about as fast as the bottom subsided. This slow subter ranean movement was varied by uprisings here and there, and notably by the outburst at successive periods of a great group of active submarine volcanoes over Wales, the Lake district, and the south of Ireland; but at the close of the Silurian period a vast series of disturbances took place, as the consequence of which the first rough outlines of the European continent were blocked out. The floor of the sea was raised into long ridges of land, among which were some on the site of the Alps, the Spanish peninsula, and the hills of the west and north of Britain. The thick mass of marine sediment was crumpled up, and here and there even converted into hard crystalline rock. Large enclosed basins, gradually cut off from the sea, like the modern Caspian and Sea of Aral, extended from beyond the west of Ireland across to Scandinavia and even into the west of Russia. These lakes abounded in bone-covered fishes of strange and now long-extinct types, while the land around was clothed with a club-moss and reed-like vegetation— *Psilophyton, Sigillaria, Calamite*, etc.—the oldest terrestrial flora of which any abundant records have yet been found in Europe. The sea, dotted with numerous islands, appears to have covered most of the heart of the continent.

A curious fact deserves to be noticed here. During the convulsions by which the sediments of the Silurian sea-floor were crumpled up, crystallised, and elevated into land, the area of Russia seems to have remained nearly unaffected. Not only so, but the same immunity from violent disturbance has prevailed over that vast territory during all subsequent geological periods. The Ural Mountains on the east have again and again served as a line of relief, and have been from time to time ridged up anew. The German domains on the west have likewise suffered extreme convulsion. But the wide intervening plateau of Russia has apparently always maintained its flatness either as sea-bottom or as terrestrial plains. As I have already remarked, there has been a remarkable persistence alike in exposure to and immunity from terrestrial disturbance. Areas that lay along lines of weakness have suffered repeatedly in successive geological revolutions, while tracts outside of these regions of convulsion have simply moved gently up or down without material plication or fracture.

By the time of the coal growths, the aspect of the European area had still further changed. It then consisted of a series of low ridges or islands in the midst of a shallow sea or of wide salt-water lagoons. A group of islands occupied the site of some of the existing high grounds of Britain. A long, irregular ridge ran across what is now France from Brittany to the Mediterranean. The Spanish peninsula stood as a detached island. The future Alps rose as a long, low ridge, to the north of the eastern edge of which lay another insular space, where now we find the high grounds of Bavaria and Bohemia. The shallow waters that wound among these scattered patches of land were gradually silted up. Many of them became marshes, crowded with a most luxuriant cryptogamic vegetation, specially of lycopods and ferns, while the dry grounds waved green with coniferous trees. By a slow intermittent subsidence, islet after islet sank beneath the verdant swamps. Each fresh depression submerged the rank jungles and buried them under sand and mud, where they were eventually compressed into coal. To this united co-operation of dense vegetable growth, accumulation of sediment, and slow subterranean movement, Europe owes her coal-fields.

All this time the chief area of high ground in Europe appears still to have lain to the north and north-west. The old gnarled gneiss of that region, though constantly worn down and furnishing materials towards each new formation, yet rose up as land. It no doubt received successive elevations during the periods of disturbance, which more or less compensated for the constant loss from its surface.

The next scene we may contemplate brings before us a series of salt lakes, covering the centre of the continent from the north of Ireland to the heart of Poland. These basins were formed by the gradual cutting off of portions of the sea which had spread over the region. Their waters were red and bitter, and singularly unfavourable to life. On the low intervening ridges a coniferous and cycadaceous vegetation grew, sometimes in quantity sufficient to supply materials for the formation of coal-seams. The largest of these salt lakes stretched from the edge of the old plateau of Central France along the base of the Alpine ridge to the high grounds of Bohemia, and included the basin of the Rhine from Bâle down to the ridge beyond Mayence, which has been subsequently cut through by the river into the picturesque gorge between Bingen and the Siebengebirge. This lake was filled up with red sand and mud, limestone, and beds of rock salt. Where the eastern Alps now rise the opener waters were the scene of a long-continued growth of dolomite, out of which in later ages the famous dolomite mountains of the Tyrol were carved.

These salt lakes of the Triassic period seem to have been everywhere quietly effaced by a widespread depression, which allowed the water of the main ocean once more to overspread the greater part of Europe. This slow subsidence went on so long as to admit of the accumulation of masses of limestone, shale, and sandstone, several thousand feet in thickness, and probably to bring most of the insular tracts of Central Europe under water. To this period, termed by geologists the Jurassic, we can trace back the origin of a large part of the rock now forming the surface of the continent, from the low plains of Central England up to the crests of the northern Alps, while in the Mediterranean basin, rocks of the same age cover a large area of the plateau of Spain, and form the central mass of the chain of the Apennines. It is interesting to know that the north-west of Britain continued still to rise as land in spite of all the geographical changes which had taken place to the south and east. We can trace even yet the shores of the Jurassic sea along the skirts of the mountains of Skye and Ross-shire.

The next long era, termed the Cretaceous, was likewise more remarkable for slow accumulation of rock under the sea than for the formation of new land. During that time the Atlantic sent its waters across the whole of Europe and into Asia. But they were probably nowhere more than a few hundred feet deep over the site of our continent, even at their deepest part. Upon their bottom there gathered a vast mass of calcareous mud, composed in great part of foraminifera, corals, echinoderms, and molluscs. Our English chalk which ranges across the north of France, Belgium, Denmark, and the North of Germany, represents a portion of the deposits of that sea-floor, probably accumulated in a northern, somewhat isolated basin, while the massive hippurite limestone of Southern Europe represents the deposits of the opener ocean. Some of the island spaces which had remained for a vast period above water, and had by their degradation supplied materials for the sediment of successive geological formations, now went down beneath the Cretaceous sea. The ancient highgrounds of Bohemia, the Alps, the Pyrenees, and the Spanish tableland were either entirely submerged, or at least had their area very considerably reduced. The submergence likewise affected the north-west of Britain; the western highlands of Scotland lay more than 1000 feet below their present level.

When we turn to the succeeding geological period, that of the Eocene, the proofs of widespread submergence are still more striking. A large part of the Old World seems to have sunk down; for we find that one wide sea extended across the whole of Central Europe and Asia. It was at the close of this period of extreme depression that those subterranean movements began to which the present configuration of Europe is mainly due. The Pyrenees, Alps, Apennines, Carpathians, the Caucasus, and the heights of Asia Minor mark, as it were, the crests of the vast earth-waves into which the solid framework of Europe was then thrown. So enormous was the contortion that, as may be seen along the northern Alps, the rocks for thousands of feet were completely inverted, this inversion being accompanied by the most colossal folding and twisting. The massive sedimentary formations were crumpled up, and doubled over each other, as we might fold a pile of cloth. In the midst of these commotions the west of Europe remained undisturbed. It is strange to reflect that the soft clays and sands under London are as old as some of the hardened rocks which have been upheaved into such picturesque peaks along the northern flanks of the Alps.

After the completion of these vast terrestrial disturbances, the outlines of Europe began distinctly to shape themselves into their present form. The Alps rose as a great mountain range, flanked on the north by a vast lake which covered all the present lowlands of Switzerland, and stretched northward across a part of the Jura Mountains, and eastward into Germany. The size of this fresh-water basin may be inferred from the fact that one portion only of the sand and gravel that accumulated in it even now measures 6000 feet in thickness. The surrounding land was densely clothed with a vegetation indicative of a much warmer climate than Europe now can boast. Palms of American types, as well as date palms, huge Californian pines (Sequoia), laurels, cypresses, and evergreen oaks, with many other evergreen trees, gave a distinctive character to the vegetation. Among the trees too were planes, poplars, maples, willows, oaks, and other ancestors of our living woods and forests; numerous ferns grew in the underwood, while clematis and vine wound themselves among the branches. The waters were haunted by huge pachyderms, such as the dinotherium and hippopotamus; while the rhinoceros and mastodon roamed through the woodlands.

A marked feature of this period in Europe was the abundance and activity of its volcanoes. In Hungary, Rhineland, and Central France, numerous vents opened and poured out their streams of lava and showers of ashes. From the south of Antrim, also, through the west coast of Scotland, the Faröe Islands, and Iceland, even far into Arctic Greenland, a vast series of fissure-eruptions poured forth successive floods of basalt, fragments of which now form the extensive volcanic plateaux of these regions.

The mild climate indicated by the vegetation in the deposits of the Swiss lake prevailed even into Polar latitudes, for the remains of numerous evergreen shrubs, oaks, maples, walnuts, hazels, and many other trees have been found in the far north of Greenland, and even within 8° 15′ of the pole. The sea still occupied much of the lowlands of Europe. Thus it ran as a strait between the Bay of Biscay and the Mediterranean, cutting off the Pyrenees and Spain from the rest of the continent. It swept round the north of

France, covering the rich fields of Touraine and the wide flats of the Netherlands. It rolled far up the plains of the Danube, and stretched thence eastward across the south of Russia into Asia.

By this time some of the species of shells which still people the European seas had appeared. So long have they been natives of our area that they have witnessed the rise of a great part of the continent. Some of the most stupendous changes which they have seen have taken place in the basin of the Mediterranean, where, at a comparatively recent geological period, parts of the sea-floor were upheaved to a height of 3000 feet. It was then that the breadth of the Italian peninsula was increased by the belt of lower hills that flanks the range of the Apennines. Then, too, Vesuvius and Etna began their eruptions. Among these later geographical events also we must place the gradual isolation of the Sea of Aral, the Caspian, and the Black Sea from the rest of the ocean, which is believed to have once spread from the Arctic regions down the west of Asia, along the base of the Ural Mountains into the south-east of Europe.

The last scene in this long history is one of the most unexpected of all. Europe, having nearly its present height and outlines, is found swathed deep in snow and ice. Scandinavia and Finland are one vast sheet of ice, that creeps down from the watershed into the Atlantic on the one side, and into the basin of the Baltic on the other. All the high grounds of Britain are similarly buried. The bed of the North Sea as well as of the Baltic is in great measure choked with ice. The Alps, the Pyrenees, the Carpathians, and the Caucasus send down vast glaciers into the plains at their base. Northern plants find their way south even to the Pyrenees, while the reindeer, muskox, lemming, and their Arctic companions, roam far and wide over France.

As a result of the prolonged passage of solid masses of ice over them, the rocks on the surface of the continent, when once more laid bare to the sun, present a worn, flowing outline. They have been hollowed into basins, ground smooth, and polished. Long mounds and wide sheets of clay, gravel, and sand have been left over the low grounds, and the hollows between them are filled with innumerable tarns and lakes. Crowds of boulders have been perched on the sides of the hills and dropped over the plains. With the advent of a milder temperature the Arctic vegetation gradually disappeared from the plains. Driven up step by step before the advancing flora from more genial climates, it retired into the mountains, and there to this day continues to maintain itself. The present Alpine flora of the Pyrenees, the Alps, Britain, and Scandinavia, is thus a living record of the ice-age. The reindeer and his friends have long since been forced to return to their northern homes.

After this long succession of physical revolutions, man appears as a denizen of the Europe thus prepared for him. The earliest records of his presence reveal him as a fisher and hunter, with rude flint-pointed spear and harpoon. And doubtless for many a dim century such was his condition. He made no more impress on external nature than one of the beasts which he chased. But in course of time, as civilisation grew, he asserted his claim to be one of the geographical forces of the globe. Not content with gathering the fruits and capturing the animals which he found needful for his wants, he gradually entered into a contest with nature to subdue the earth and to possess it. Nowhere has this warfare been fought out so vigorously as on the surface of Europe. On the one hand, wide dark regions of ancient forest have given place to smiling cornfields. Peat and moor have made way for pasture and tillage. On the other hand, by the clearance of woodlands the rainfall has been so diminished that drought and barrenness have spread where verdure and luxuriance once prevailed. Rivers have been straitened and made to keep their channels, the sea has been barred back from its former shores. For many generations the surface of the continent has been covered with roads, villages and towns, bridges, aqueducts and canals, to which this century has added a multitudinous network of railways, with their embankments and tunnels. In short, wherever man has lived, the ground beneath him bears witness to his presence. It is slowly covered with a stratum either wholly formed by him or due in great measure to his operations. The soil under old cities has been increased to a depth of many feet by the rubbish of his buildings; the level of the streets of modern Rome stands high above that of the pavements of the Cæsars, and that again above the roadways of the early Republic. Over cultivated fields his potsherds are turned up in abundance by the plough. The loam has risen within the walls of his graveyards as generation after generation has mouldered into dust.

It must be owned that man, in much of his struggle with the world around him, has fought blindly for his own ultimate interests. His contest, successful for the moment, has too often led to sure and sad disaster. Stripping forests from hill and mountain, he has gained his immediate object in the possession of their abundant stores of timber; but he has laid open the slopes to be parched by drought, or swept bare by rain. Countries once rich in beauty, and plenteous in all that was needful for his support, are now burnt and barren, or almost denuded of their soil. Gradually he has been taught, by his own bitter experience, that while his aim still is to subdue the earth, he can attain it, not by setting nature and her laws at defiance, but by enlisting them in his service. He has learnt at last to be the minister and interpreter of nature, and he finds in her a ready and unrepining slave.

In fine, looking back across the long cycles of change through which the land has been shaped into its present form, let us realise that these geographical revolutions are not events wholly of the dim past, but that they are still in progress. So slow and measured has been their march, that even from the earliest times of human history they seem hardly to have advanced at all. But none the less are they surely and steadily transpiring around us. In the fall of rain and the flow of rivers, in the bubble of springs and the silence of frost, in the quiet creep of glaciers and the tumultuous rush of ocean waves, in the tremor of the earthquake and the outburst of the volcano, we may recognise the same play of terrestrial forces by which the framework of the continents has been step by step evolved. In this light the familiar phenomena of our daily experience acquire an historical interest and dignity. Through them we are enabled to bring the remote past vividly before us, and to look forward hopefully to that great future in which, in the physical not less than in the moral world, man is to be a fellow-worker with God.

XIV.

THE GEOLOGICAL INFLUENCES WHICH HAVE AFFECTED THE COURSE OF BRITISH HISTORY.[1]

PROBABLY few readers realise to how large an extent the events of history have been influenced by the geological structure of the ground whereon they have been enacted. I propose to illustrate this influence from some of the more salient features in the early human occupation of the British Islands, and in the subsequent historical progress of the English people. No better proof of the reality of the relation in question could be given than the familiar contrast between the heart of England and the heart of Scotland. The one area is a region of low plains, inhabited by an English-speaking race; richly agricultural in one part, teeming with a busy mining population in another; dotted with large cities; the air often foul from the smoke of thousands of chimneys, and resonant with the clanking of innumerable manufactories and the screams of locomotives flying hither and thither over a network of railways. The other region is one of rugged mountains and narrow glens

---

[1] Introductory Lecture at the opening of the session of the Class of Geology in the University of Edinburgh, November 1881. *Macmillan's Magazine*, March 1882.

tenanted by a Celtic race that, keeping to its old Gaelic tongue and primitive habits, has never built towns, hardly even villages—a region partly devoted to pasture and still haunted by the game and wild animals of primeval times, but with no industrial centres, no manufactures of any kind, and only a feeble agriculture that struggles for existence along the bottoms of the valleys. Now, why should two parts of the same small country differ so widely from each other? To give a complete answer to the question would of course involve a detailed examination of the history of each area. But we should find that fundamentally the differences have arisen from the originally utterly distinct geological structure of the two regions. This diversity of structure initiated the divergences in human characteristics even in far prehistoric times, and it continues, even in spite of the blending influences of modern civilisation, to maintain them down to the present day.

Let us first briefly consider what was the probable condition of Britain at the time when the earliest human beings appeared in the country. At that ancient epoch there can be no doubt that the British Islands still formed part of the mainland of Continental Europe. There is reason to believe that the general level of these islands may have been then considerably higher than it has been since. From the shape of the bottom of the Atlantic immediately to the west of our area, as revealed by the abundant soundings and dredgings of recent years, it is evident that if the British Islands were now raised even 1000 feet or more above their present level, they would not thereby gain more than a belt of lowland somewhere about 200 miles broad on their western border. They stand, in fact, nearly upon the edge of the great European plateau which, about 230 miles to the west of them, plunges rapidly down into the abysses of the Atlantic. It is perfectly certain, therefore, that though our area was formerly prolonged westward beyond its present limits, there has never been any important mass of land to the west of us in recent geological times, or within what we call the human period, probably never at any geological epoch at all. Every successive wave of migration, whether of plant or of animal, must have come from the other or eastern side. But though our country could never have stretched much beyond its present westward limits, it once undoubtedly spread eastward over the site of what is now the North Sea. Even at the present day, an elevation of less than 600 feet would convert the whole of that sea into dry land from the north of Shetland to the headlands of Brittany. At the time when these wide plains united Britain to the mainland, the Thames was no doubt a tributary of the Rhine, which, in its course northward, may have received other affluents from the east of Britain before it poured its waters into the Atlantic somewhere between the heights of Shetland and the mountainous coasts of Southern Norway.

There is evidence of remarkable oscillations of climate at the epoch of the advent of man into this part of Europe. A time of intense cold, known as the Ice Age or Glacial period, was drawing to a close. Its glaciers, frozen rivers and lakes, and floating icebergs, had converted most of Britain, and the whole of Northern Europe, into a waste of ice and snow, such as North Greenland still is; but the height of the cold was past, and there now came intervals of milder seasons, when the wintry mantle was withdrawn northward, so as to allow the vegetation and the roaming animals of more temperate latitudes to spread westward into Britain. From time to time a renewal of the cold once more sent down the glaciers into the valleys, or even into the sea, froze the rivers over in winter, and allowed the Arctic flora and fauna again to migrate southward into tracts from which the temperate plants and animals were forced by the increasing cold to retreat. At last, however, the Arctic conditions of climate ceased to reappear, and the Arctic vegetation, with its accompanying reindeer, musk-sheep, lemming, Arctic fox, glutton, and other northern animals, retreated from our low grounds. Of these ancient chilly periods, however, the Arctic plants still found on our mountain tops remain as living witnesses, for they are doubtless descendants of the northern vegetation which overspread Britain when still part of the continent, and before the arrival of our present temperate flora and fauna.

Previous to the final retreat of the ice, the alternating warmer intervals brought into Britain many wild animals from wilder regions to the south. Horses, stags, Irish elks, roe deer, wild oxen, and bisons roamed over the plains; wild boars, three kinds of rhinoceros, two kinds of elephant, brown bears and grizzly bears, haunted the forests. The rivers were tenanted by the hippopotamus, beaver, otter, water-rat; while among the carnivora were wolves, foxes, wild cats, hyænas, and lions. Many of these animals must have moved in herds across the plains, over which the North Sea now rolls. Their bones have been dredged up in hundreds by the fishermen from the surface of the Dogger-Bank.

Such were the denizens of southern England when man made his first appearance there. It seems not unlikely that he came some time before the close of the long Ice Age. He may have been temporarily driven out of the country by the returning cold periods, but would find his way back as the climate ameliorated. Much ingenuity has been expended in tracing a succession of civilisation in this primeval human population of Britain. Among the records of its presence there have been supposed to be traces of an earlier race of hunters of a low order, furnished with the rudest possible stone implements; and a later people, who, out of the bones of the animals they captured, supplied themselves with deftly-made, and even artistically-decorated weapons. All that seems safely deducible from the evidence, however, may be summed up in saying that the *palæolithic* men, or men of the older stone period, who hunted over the plains, and fished in the rivers, and lived in the caves of this country, have left behind them implements, rude indeed, but no doubt quite suitable for their purpose; and likewise other weapons and tools of a more finished kind, which bear a close relationship to the implements still in use among the modern Eskimos. It has been suggested that the Eskimos are their direct descendants, driven into the inhospitable north by the pressure of more warlike races.

The rude hunter and dweller in caves passed away before the advent of the farmer and herdsman of the *Neolithic* or later stone period. We know much more of him than of his predecessors. He was short of stature, with an oblong head, and probably a dark skin and dark curly hair. His implements of stone were often artistically fashioned and polished. Though still a hunter and fisher, he knew also how to farm. He had flocks and herds of domestic animals; he was acquainted with the arts of spinning and weaving, could make a rude kind of pottery, and excavate holes and subterranean galleries in the chalk for the extraction of flints for his weapons and tools. That he had some notion of a future state may be inferred from arrow-heads pottery, and implements of various kinds which are found in his graves, evidently placed there for the use of the departed. He has been regarded as probably of a Non-Aryan race, of which perhaps the modern Basques are lineal descendants, isolated among the fastnesses of the Pyrenees by the advance of younger tribes. Traces of his former presence in Britain have been conjectured to be recognisable in the small dark Welshmen, and the short swarthy Irishmen of the west of Ireland.

When the earliest Neolithic men appeared in this region, Britain may have still been united to the continent. But the connection was eventually broken. It is obvious that no event in the geological history of Britain can have had a more powerful influence on its human history than the separation of the country as a group of islands cut off by a considerable channel from direct communication with the mainland of Europe. Let us consider for a moment how the disconnection was probably brought about.

There can be no doubt that at the time when Britain became an island, the general contour of the country was, on the whole, what it is still. The same groups of mountains rose above the same plains and valleys, which were traversed by the same winding rivers. We know that in the glacial and later periods considerable oscillations of level took place; for, on the one hand, beds of sea-shells are found at heights of 1200 or 1300 feet above the present sea-level; and, on the other hand, ancient forest-covered soils are now seen below tide-mark. It was doubtless mainly subsidence that produced the isolation of Britain. The whole area slowly sank, until the lower tracts were submerged, the last low ridge connecting the land with France was overflowed, and Britain became a group of islands. But unquestionably the isolation was helped by the ceaseless wear and tear of the superficial agencies which are still busy at the same task. The slow but sure washing of descending rain, the erosion of watercourses, and the gnawing of sea-waves, all told in the long degradation. And thus, foundering from want of support below, and eaten away by attacks above, the low lands gradually diminished, and disappeared beneath the sea.

Now, in this process of separation, Ireland unfortunately became detached from Britain. We have had ample occasion in recent years to observe how much this geological change has affected our domestic history. That the isolation of Ireland took place before Britain had been separated from the continent may be inferred from a comparison of the distribution of living plants and animals. Of course, the interval which had then elapsed since the submergences and ice-sheets of the glacial period must have been of prodigious duration, if measured by ordinary human standards. Yet it was too short to enable the plants and animals of Central Europe completely to possess themselves of the British area. Generation after generation they were moving westward, but long before they could all reach the north-western seaboard, Ireland had become an island, so that their further march in that direction was arrested, and before the subsequent advancing bands had come as far as Britain, it too had been separated by a sea channel which finally barred their progress. Comparing the total land mammals of the west of Europe, we find that while Germany has ninety species, Britain has forty, and Ireland only twenty-two. The reptiles and amphibia of Germany number twenty-two, those of Britain thirteen, and those of Ireland four. Again, even among the winged tribes, where the capacity for dispersal is so much greater, Britain possesses twelve species of bats, while Ireland has no more than seven, and 130 land birds to 110 in Ireland. The same discrepancy is traceable in the flora, for while the total number of species of flowering plants and ferns found in Britain amounts to 1425, those of Ireland number 970—about two-thirds of the British flora. Such facts as these are not explicable by any difference of climate rendering Ireland less fit for the reception of more varied vegetation and animal life; for the climate of Ireland is really more equable and genial than that of the regions lying to the east of it. They receive a natural and consistent interpretation on the assumption of the gradual separation of the British Islands during a continuous north-westward migration of the present flora and fauna from Central Europe.

The last neck of land which united Britain to the mainland was probably that through which the Strait of Dover now runs. Apart from the general subsidence of the whole North Sea area, which is attested by submerged forests on both sides, it is not difficult to perceive how greatly the widening of the channel has been aided by waves and tidal currents. The cliffs of Kent on the one side and of the Boulonnais on the other, ceaselessly battered by the sea, and sapped by the trickle of percolating springs, are crumbling before our very eyes. The scour of the strong tides which pour alternately up and down the strait, must have helped also to deepen the Channel. And yet, in spite of the subsidence and this constant erosion, the depression remains so shallow that its deepest parts are less than 180 feet below the surface. As has often been remarked, if St. Paul's Cathedral could be shifted from the heart of London to the middle of the strait more than half of it would rise above water.

At what relative time in the human occupation of the region this channel was finally opened cannot be determined. At first the strait was doubtless much narrower than it has since become, so that it would not oppose the same obstacle to free intercourse which it now does, and Neolithic man may have readily traversed it in his light coracle of skins. Be this as it may, there can be no doubt that the old Basque or Iberian stock had for many ages inhabited Britain before the succeeding wave of human migration advanced to overflow and efface it. The next invaders—the first advance-guard of the great Aryan family —were Celts, whose descendants still form a considerable part of the population of the British Isles. The Celt differed in many respects from the small swarthy Iberian whom he supplanted. He was tall, round-headed, and fair-skinned, with red or brown hair. Endowed with greater bodily strength and pugnacity, he drove before him the older smaller race of short oblong-headed men, gradually extirpating them, or leaving here and there, in less attractive portions of the country, small island-like remnants of them which insensibly mingled with their conquerors, though, as I have already remarked, traces of these remnants are perhaps partially recognisable in the characteristic Iberian-like lineaments of some districts of the country even at the present day.

The Celts, as we now find them in Britain, belong to two distinct divisions of the race, the Irish or Gaelic, and the Welsh or Cymric. Some difference of opinion has arisen as to which of these branches appeared in the country first. It seems to me that if the question is discussed on the evidence of geological analogy, the unquestionable priority should be assigned to the Gaels. There can be no doubt that the Celts came from the east. They had already overspread Gaul and Belgium before they invaded Britain. The tribe which is found on the most northerly and westerly tracts should be the older, having crossed, on its way, the regions lying to the east, while on the other hand, the race occupying the eastern tracts should be of later origin. We ought to judge of the spread of the human population as we do of that of the flora and fauna. Had England been already occupied by the Welsh, Cymric or British branch, it is inconceivable that the Irish or Gaelic branch could have marched through the territory so occupied, and have established itself in Scotland and Ireland. The Gaels were, no doubt, the first to arrive. Finding the country inhabited by the little Neolithic folk they dispossessed them, and spread by degrees over the whole of the islands. At a later time the Cymry arose. We are not here concerned with the question whether these originated by a gradual bifurcation in the development of the Celtic race after its settlement within Britain, or came as a later Celtic wave of migration from the continent. It is enough to notice that they are found at the beginning of the historical period to be in possession of England, Wales, and the south of Scotland up to the estuary of the Clyde. It is improbable that the Gaels, who no doubt once occupied the same attractive region, would have willingly quitted it for the more inhospitable moors of Scotland and the distant bogs and fenlands of Ireland. It is much more likely that they were driven forcibly out of it. Possibly the traditions they carried with them of the greater fertility of England may have instigated the numerous inroads which from early Roman times downwards they made to recover the lands of their forefathers. Crossing from Ireland they repossessed themselves of the west of Wales, and sweeping down from the Scottish Highlands they repeatedly burst across the Roman wall, carrying pillage and rapine far into the province where their Cymric cousins had begun to learn some of the arts and effeminacy of Roman civilisation.

Looking at the territory occupied by the Cymry at the time of their greatest extension, we can see how their course northward was influenced by geological structure. As they advanced along the plains which lay on the west side of the great Pennine chain of the centre and north of England, they encountered the range of fells which connects the mountain group of Cumberland and Westmoreland with the uplands of Yorkshire and Durham. This would probably be for some time a barrier to their progress. But after crossing it by some of the deep valleys by which it is trenched, they would find themselves in the wide plains of the Eden and the Solway. Still pushing their way northward, and driving the Gaels before them, they would naturally follow the valley of the Nith, leaving on the left hand the wild mountainous region of Galloway, or "country of the Gael," to which the conquered tribe retired, and on the right the high moorlands about the head of Clydesdale and Tweeddale. Emerging at last upon the lowlands of Ayrshire and lower Clydesdale, they would spread over them until their further march was arrested by the great line of the Highland mountains. Into these fastnesses, stoutly defended by the Pictish Gaels, they seem never to have penetrated. But they built, as their northern outpost, the city and castle of Alcluyd, where the picturesque rock of Dumbarton, or " fort of the Britons," towers above the Clyde.

At one time, therefore, the Cymry extended from the mouth of the Clyde to the south of England. One language—Welsh and its dialects—appears to have been spoken throughout that territory. Hence the battles of King Arthur, which, from the evidence of the ancient Welsh poems, appear to have been fought, not in the south-west of England, as is usually supposed, but in the

middle of Scotland, against the fierce Gwyddyl Ffichti or
Picts of the north and the heathen swarming from beyond the sea, were sung all the way down into Wales and
Devon, and across the Channel among the vales of Brittany,
whence, becoming with every generation more mystical and
marvellous, they grew into favourite themes of the romantic
poetry of Europe.

The Roman occupation affected chiefly the lowlands of
England and Scotland where the more recent geological
formations extend in broad plains or plateaux. Numerous
towns were built there, between which splendid roads
extended across the country. The British inhabitants of
these lowlands were not extirpated, but continued to live
on the lands which they had tilled of old, more or less
affected by the Roman civilisation with which, for some
four centuries or more, they were brought in contact. But
the regions occupied by the more ancient rocks, rising
into rugged forest-covered mountains, offered an effective
barrier to the march of the Roman legions, and afforded
a shelter within which the natives could preserve their
ancient manners and language with but little change. The
Romans occupied the broad central lowland region of
Scotland which is formed by the Old Red Sandstone and
Carboniferous strata, extending up to the base of the
Highlands. But though they inflicted severe defeats upon
the wild barbarians who issued from the dark glens, and
though they seem to have been led by Severus round by
the Aberdeenshire low grounds to the shores of the Moray
Firth, and to have returned through the heart of the Highlands, they were never able permanently to bring any part
of the mountainous area of crystalline rocks under their
rule.

The same geological influences which guided the progress of the Roman armies may be traced in the subsequent Teutonic invasions of Angles, Saxons, Jutes, and Norwegians. Arriving from the east and north-east, these hordes found level lowlands open to their attack. Where no impenetrable thicket, forest, fenland, or mountainous barrier impeded their advance, they rapidly pushed inland, utterly extirpating the British population and driving its remnants steadily westward. By the end of the sixth century the Britons had disappeared from the eastern half of the island south of the Firth of Forth. Their frontier, everywhere obstinately defended, was very unequal in its capabilities of defence. In the north, where they had been driven across bare moors and bleak uplands, they found these inhospitable tracts for a time a barrier to the further advance of the enemy; but where they stood face to face with their foe in the plains they could not permanently resist his advance. This difference in physical contour and geological structure led to the final disruption of the Cymric tract of country by the two most memorable battles in the early history of England.

Between the Britons of South Wales and those of Devon and Cornwall lay the rich vale of the Severn. Across this plain there once spread in ancient geological times a thick sheet of Jurassic strata of which the bold escarpment of the Cotswold Hills forms a remnant. The valley has been in the course of ages hollowed out of these rocks, the depth of which is only partly represented by the height of the Cotswold plateau. The Romans had found their way into this fertile plain, and, attracted by the hot springs which still rise there, had built the venerable city of Bath and other towns. One hundred and seven years after the Romans quitted Britain, the West Saxons, who had gradually pushed their way westward up the valley of the Thames, found themselves on the edge of the Cotswold plateau, looking down upon the rich and long settled plains of the Severn. Descending from these heights, they fought in 577 the decisive battle of Deorham, which had the effect of giving them possession of the Severn valley, and thus of isolating the Britons of Devon and Cornwall from the rest of their kinsmen. Driven thus into the south-west corner of England upon ancient Devonian and granitic rocks, poorer in soil, but rich in wealth of tin and copper, these Britons maintained their individuality for many centuries. Though they have now gradually been fused into the surrounding English-speaking people, it was only about the middle of last century that they ceased to use their ancient Celtic tongue.

Still more important was the advance of the Angles on the north side of Wales. The older Palæozoic rocks of the principality form a mass of high grounds which, flanked with a belt of coal-bearing strata, descend into the plains of Cheshire. Younger formations of soft red Triassic marl and sandstone stretch northward to the base of the Carboniferous and Silurian hills of north Lancashire. This strip of level and fertile ground, bounded on the eastern side by high desert moors and impenetrable forests, connected the Britons of Wales with those of the Cumbrian uplands, and, for nearly 200 years after the Romans had left Britain, was subject to no foreign invasion, save perhaps occasional piratical descents from the Irish coasts. But at last, in the year 607, the Angles, who had overspread the whole region from the Firth of Forth to the south of Suffolk, crossed the fastnesses of the Pennine Chain and burst upon the inhabitants of the plains of the Dee. A great battle was fought at Chester in which the Britons were routed. The Angles obtained permanent possession of these lowlands, and thus the Welsh were effectually cut off from the Britons of Cumbria and Strathclyde. The latter have gradually mingled with their Teutonic neighbours, though the names of many a hill and river bear witness to their former sway. The Welsh, on the other hand, driven into their hilly and mountainous tracts of ancient Palæozoic rocks, have maintained their separate language and customs down to the present day.

Turning now to the conflict between the Celtic and Teutonic races in Scotland, we notice in how marked a manner it was directed by the geological structure of the country. The level Secondary formations which, underlying the plains, form so notable a feature in the scenery of England, are almost wholly absent from Scotland. The Palæozoic rocks of the latter kingdom have been so crumpled and broken, so invaded by intrusions of igneous matter from below, and over two-thirds of the country rendered so crystalline and massive, that they stand up for the most part as high tablelands, deeply trenched by narrow valleys. Only along the central counties between the base of the Highlands on the one side and the southern uplands on the other, where younger Palæozoic formations occur, are there any considerable tracts of lowland, and even these are everywhere interrupted by protrusions of igneous rock forming minor groups of hills or isolated crags like those that form so characteristic a feature in the landscapes around Edinburgh. In old times dense forests and impenetrable morasses covered much of the land. A country fashioned and clothed in this manner is much more suitable for defence than for attack. The high mountainous interior of the north, composed of the more ancient crystalline rocks, which had sheltered the Caledonian tribes from the well-ordered advance of the Roman legions, now equally protected them from the sudden swoop of Saxon and Scandinavian sea-pirates. Neither Roman nor Teuton ever made any lasting conquest of that territory. It has remained in the hands of its Celtic conquerors till the present time.

But the case has been otherwise with the tracts where the younger Palæozoic deposits spread out from the base of the Highland mountains. These strata have not partaken of the violent corrugations and marked crystallisation to which the older rocks have been subjected. On the contrary, they extend in gentle undulations forming level plains, and strips of lowland between the foot of the more ancient hills and the margin of the sea. It was on these platforms of undisturbed strata that invaders could most successfully establish themselves. So dominant has been this geological influence, that the line of boundary between the crystalline rocks and the Old Red Sandstone, from the north of Caithness to the coast of Kincardineshire, was almost precisely that of the frontier established between the old Celtic natives and the later hordes of Danes and Northmen. To this day, in spite of the inevitable commingling of the races, it still serves to define the respective areas of the Gaelic-speaking and English-speaking populations. On the Old Red Sandstone we hear only English, often with a northern accent, and even with not a few northern words that seem to remind us of the Norse blood which flows in the veins of these hardy fisher-folk and farmers. We meet with groups of villages and towns; the houses, though often poor and dirty, are for the most part solidly built of hewn stone and mortar, with well-made roofs of thatch, slate, or flagstone. The fuel in ordinary use is coal brought by sea from the south. But no sooner do we penetrate within the area of the crystalline rocks than all appears changed. Gaelic is now the vernacular tongue. There are few or no villages. The houses, built of boulders gathered from the soil and held together with mere clay or earth, are covered with frail roofs of ferns, straw, or heather, kept down by stone-weighted ropes of the same materials. Fireplaces and chimneys are not always present, and the pungent blue smoke from fires of peat or turf finds its way out by door and window, or beneath the begrimed rafters. The geological contrast of structure and scenery which allowed the Teutonic invaders to drive the older Celtic people from the coast-line, but prevented them from advancing inland, has sufficed during all the subsequent centuries to keep the two races apart.

On the north-western coasts of the island there are none of the fringes of more recent formations which have had so marked an influence on the east side. From the north of Sutherland to the headlands of Argyle the more ancient rocks of the country rise steep and rugged out of the sea, projecting in long bare promontories, for ever washed by the restless surge of the Atlantic. Here and there the coast-line sinks into a sheltered bay, or is interrupted by some long winding inlet that admits the ebb and flow of the ocean tides far into the heart of the mountains. Only in such depressions could a seafaring people find safe harbours and fix their settlements. When the Norsemen sailed round the north-west of Scotland they found there the counterpart of their own native country—the same type of bare, rocky, island-fringed coast-line sweeping up into bleak mountains, winding into long sea-lochs or fjords beneath the shadow of sombre pine-forests, and to the west the familiar sweep of the same wide blue ocean. So striking even now is this resemblance, that the Scot who for the first time sails along the western seaboard of

Norway can hardly realise that he is not skirting the coast-line of Inverness, Ross, or Sutherland. Such a form of coast forbade easy communication by land between valley and valley. Detached settlements arose in the more sheltered bays, where glens, opening inland, afforded ground for tillage and pasture. But the intercourse between them would be almost wholly by boat, for there could be no continuous line of farms, villages, and roads like those for which the Old Red Sandstone selvages afforded such facilities on the eastern coast. Hence, though the Norsemen possessed themselves of every available bay and inlet, driving the Celts into the more barren interior, the natural contours made it impossible that their hold of the ground should be so firm as that of their kinsmen in the east. When that hold began to relax, the Gaelic natives of the glens came down once more to the sea, and all obvious trace of the Norse occupation eventually disappeared, save in the names given by the sea-rovers to the islands, promontories, and inlets — the "ays," "nishes," or "nesses," and "fords" or fjords — which, having been adopted by the Celtic natives, show that there must have been some communication and probable intermarriage between the races. Among the outer islands the effects of the Norwegian occupation were naturally more enduring, though even there the Celtic race has long recovered its ground. Only in the Orkney and Shetland group have the Vikings left upon the physical frame and the language of the people the strong impress of their former presence. To this day a Shetlander speaks of going to Scotland, meaning the mainland, much as a Lowland Scot might talk of visiting England, or an Englishman of crossing to Ireland.

But besides governing in no small degree the distribution of races in Britain, the geological structure of the country has probably not been without its influence upon the temperament of the people. Let us take the case of the Celts, originally one great race, with no doubt the same average type of mental and moral disposition, as they unquestionably possessed the same general build of body and cast of features. Probably nowhere within our region have they remained unmixed with a foreign element, which, together with the varying political conditions under which they have lived, must have distinctly affected their character. But after every allowance has been made for these several influences, it seems to me that there are residual differences which cannot be explained except by the effects of environment. The Celt of Ireland and of the Scottish Highlands was originally the same being; he crossed freely from country to country; his language, manners and customs, arts, religion, were the same on both sides of the channel, yet no two natives of the British Islands are now marked by more characteristic differences. The Irishman seems to have changed less than the Highlander; he has retained the light-hearted gaiety, wit, impulsiveness and excitability, together with that want of dogged resolution and that indifference to the stern necessities of duty which we regard as pre-eminently typical of the Celtic temperament. The Highlander, on the other hand, cannot be called either merry or witty; he is rather of a self-restrained, reserved, unexpansive, and even perhaps somewhat sullen disposition. His music partakes of the melancholy cadence of the winds that sigh through his lonely glens; his religion, too, one of the strongest and noblest features of his character, retains still much of the gloomy tone of a bygone time. Yet he is courteous, dutiful, determinedly persevering, unflinching as a foe, unwearied as a friend, fitted alike to follow with soldier-like obedience, and to lead with courage, skill, and energy—a man who has done much in every climate to sustain and expand the reputation of the British Empire.

Now, what has led to so decided a contrast? I cannot help thinking that one fundamental cause is to be traced to the great difference between the geological structure and consequent scenery of Ireland and the Highlands. By far the greater part of Ireland is occupied by the Carboniferous limestone, which, in gently undulating sheets, spreads out as a vast plain. Round the margin of this plain the older formations rise as a broken ring of high ground, while here and there from the surface of the plain itself they tower into isolated hills or hilly groups; but there is no extensive area of mountains. The soil is generally sufficiently fertile, the climate soft, and the limestone plains are carpeted with that rich verdant pasture which has suggested the name of the Emerald Isle. In such a region, so long as the people are left free from foreign interference, there can be but little to mar the gay, careless, childlike temperament of the Celtic nature. If the country yields no vast wealth, it yet can furnish with but little labour all the necessaries of live. The Irishman is naturally attached to his holding. His fathers for generations past have cultivated the same little plots. He sees no reason why he should try to be better than they, and he resents, as an injury never to be forgiven, the attempt to remove him to where he may elsewhere improve his fortunes. The Highlander, on the other hand, has no such broad fertile plains around him. Placed in a glen, separated from his neighbours in the next glens by high ranges of rugged hills, he finds a soil scant and stony, a climate wet, cold, and uncertain. He has to fight with the elements a never-ending battle, wherein he is often the loser. The dark mountains that frown above him gather around their summits the cloudy screen which keeps the sun from ripening his miserable patch of corn, or rots it with perpetual rains after it has been painfully cut. He stands among the mountains face to face with Nature in her wilder moods. Storm and tempest, mist-wreath and whirlwind, the roar of waterfalls, the rush of swollen streams, the crash of loosened landslips, which he may seem hardly to notice, do not pass without bringing, unconsciously perhaps, to his imagination their ministry of terror. Hence the playful mirthfulness and light-hearted ease of the Celtic temperament have in his case been curdled into a stubbornness which may be stolid obstinacy or undaunted perseverance, according to the circumstances which develop it. Like his own granitic hills he has grown hard and enduring, not without a tinge of melancholy suggestive of the sadness that lingers among his wind-swept glens, and that hangs about the slopes of birk round the quiet waters of his lonely lakes. The difference between Irishman and Scot thus somewhat resembles, though on a minor scale, that between the Celt of lowland France and the Celt of the Swiss Alps, and the cause of the difference is doubtless traceable in great measure to a similar kind of contrast in their respective surroundings.

If now we turn to the influences which have been at work in the distribution of the population of the country and the development of the national industries, we find them in large degree of a geological kind.

In the first place, the feral ground, or territory left in a state of nature and given up to game, lies mostly upon rocks which, protruding almost everywhere to the surface and only scantily and sparsely covered with a poor soil, are naturally incapable of cultivation. The crystalline formations of the Scottish Highlands may be taken as an example of this kind of territory. The grouse-moors and deer-forests of that region exist there not merely because the proprietors of the land have so willed it, but because over hundreds of square miles the ground itself could be turned to no better use, for it can neither be tilled nor pastured. Much patriotic nonsense has been written about the enormity of retaining so much land as game preserves. But in this, as in so many other matters, man must be content to be the servant of Nature. He cannot plant crops where she has appointed that they shall never grow; nor can he pasture flocks of sheep where she has decreed that only the fox, the wild cat, and the eagle shall find a home.

In the second place, the true pasture lands—that is, the tracts which are too high or sterile for cultivation, but which are not too rocky to refuse to yield, when their heathy covering is burnt off, a sweet grassy herbage, excellent for sheep and cattle—lie mainly on elevated areas of non-crystalline Palæozoic rocks. The long range of pastoral uplands in the South of Scotland, and the fells of Cumberland, Northumberland, and Yorkshire, are good examples. These lonely wilds might be grouped into districts each marked off by certain distinctive types of geological structure, and consequently of scenery. And it might, for aught I know, be possible to show that these distinctions have not been without their influence upon the generations of shepherds who have spent their solitary lives among them; that in character, legends, superstitions, song, the peasants of Lammermuir might be distinguished from those of Liddesdale, and both from those of Cumberland and Yorkshire—the distinction, subtle perhaps and hardly definable, pointing more or less clearly to the contrasts between their respective surroundings.

In the third place, the sites of towns and villages may often be traced to a guiding geological influence. Going back to feudal times, we at once observe to what a large extent the positions of the castles of the nobles were determined by the form of the ground, and notably by the prominence of some crag which, rising well above the rest of the country, commanded a wide view and was capable of defence. Across the lowlands of Scotland such crags are abundantly scattered. They consist for the most part of hard projections of igneous rock, from which the softer sandstones and shales, that once surrounded and covered them, have been worn away. Many of them are crowned with mediæval fortresses, some of which stand out among the most famous spots in the history of the country. Dumbarton, Stirling, Blackness, Edinburgh, Tantallon, Dunbar, the Bass, are familiar names in the stormy annals of Scotland. A strong castle naturally gathered around its walls the peasantry of the neighbourhood for protection against the common foe, and thus by degrees the original collection of wooden booths or stone huts grew into a village or even into a populous town. The Scottish metropolis undoubtedly owes its existence in this way to the bold crag of basalt on which its ancient castle stands.

In more recent times the development of the mining industries of the country has powerfully affected both the growth and decay of towns. Comparing in this respect the maps of to-day with those of 150 or 200 years ago, we cannot but be struck with the remarkable changes that have taken place in the interval. Some places which were then of but minor importance have now advanced to the first rank, while others that were among the chief towns of the realm have either hardly advanced at all or have positively declined. If now we turn to a geological map, we find that in almost all cases the growth has taken place within or near to some important mineral field, while the decadence occurs in tracts where there are no workable minerals. Look, for example, at the prodigious increase of such towns as Glasgow, Liverpool, Manchester, Newcastle, Birmingham, and Middlesborough. Each of these owes its advance in population and wealth to its position in the midst of, or close to, fields of coal and iron. Contrast, on the other hand, the sleepy, quiet, unprogressive content, and even sometimes unmistakable decay, of not a few county towns in our agricultural districts.

Closely connected with this subject is the remarkable transference of population which for the last generation or two has been in such rapid progress among us. The large manufacturing towns are increasing at the expense of the rural districts. The general distribution of the population is changing, and the change is obviously underlaid by a geological cause. People are drawn to the districts where they can obtain most employment and best pay; and these districts are necessarily those where coal and iron can be obtained, without which no branch of our manufacturing industry could at present exist.

In the fourth place the style of architecture in different districts is largely dependent upon the character of their geology. The mere presence or absence of building-stone creates at once a fundamental distinction. Hence the contrast between the brickwork of England, where building-stone is less common, and the stonework of Scotland, where stone abounds. But even as we move from one part of a stone-using region to another, marked varieties of style may be observed, according to local geological development. The massive yellow limestone blocks of Bath or Portland, the thin blue flags and slates of the Lake district, the thick courses of deep red freestone in Dumfriesshire, the bands of fine, easily-dressed white sandstone of Edinburgh, have all produced certain differences of style and treatment. To a geological eye that passes rapidly through a territory, this character of its buildings is often suggestive of its geology.

In the fifth and last place, the dominant influence of the geology of a country upon its human progress is nowhere more marvellously exhibited than in the growth of British commerce. The internal trade of this country may be spoken of as its life-blood, pulsating unceasingly along a network of railways. This vast organism possesses not one but many hearts, from each of which a vigorous circulation proceeds. Each of these hearts or nerve-centres is situated on or near a mineral region, whence its nourishment comes. The history of the development of our system of railways, our steam machinery, our manufactures, is unintelligible except when taken together with the opening up of our resources in coal and iron.

The growth of the foreign commerce of the country enforces the same lesson. Even, however, before the days of steam navigation, her geological structure gave England a distinct advantage over her neighbours on the Continent. Owing to the denudation that has hollowed out the surface of the country, and the subsidence that has depressed the shoreward tracts beneath the sea, the coast-line of Britain abounds in admirable natural harbours, which on the opposite side of the Channel and North Sea are hardly to be found. There can be no question that in the infancy of navigation this gave a superiority for which hardly anything else could compensate. We boast that it is our insular position and our English blood that have made us sailors. Let us remember that, in spite of their less favourable position, our neighbours on the opposite shores of the Continent have become excellent sailors too, and that if we have been enabled to lead the van in international commerce it has been largely due to the abundant, safe, and commodious inlets in our coast-line which have sheltered our marine.

Of the foreign trade of the country it is not needful to speak. Its rapid growth during the present century is distinctly traceable to the introduction of steam navigation, and therefore directly to the development of those mineral resources which form so marked an element in the fortunate geological construction of the British Islands.

THE END.